ASCE Manuals and Reports on Engineering Practice No. 106

Horizontal Auger Boring Projects

Second Edition

Prepared by the
Horizontal Auger Boring Task Force of the
Trenchless Installation of Pipelines Committee of the
Utility Engineering and Surveying Institute of the
American Society of Civil Engineers

Edited by
Alan Atalah, Ph.D., P.E.
Lameck Onsarigo, Ph.D.

ASCE AMERICAN SOCIETY OF CIVIL ENGINEERS

UESI UTILITY ENGINEERING & SURVEYING INSTITUTE

Published by the American Society of Civil Engineers

Library of Congress Cataloging-in-Publication Data

Names: American Society of Civil Engineers. Horizontal Auger Boring Task Force, author. | Atalah, Alan, editor.
Title: Horizontal auger boring projects / prepared by the Horizontal Auger Boring Task Force of the Trenchless Installation of Pipelines Committee of the Utilities Engineering and Surveying Institute of the American Society of Civil Engineers, Task Committee Chair, Alan Atalah, Ph.D., P.E.
Description: Second edition. | Reston, Virginia : ASCE, UESI, [2017] | Includes bibliographical references and index. | Description based on print version record and CIP data provided by publisher; resource not viewed.
Identifiers: LCCN 2016044510 (print) | LCCN 2016042035 (ebook) | ISBN 9780784480236 (pdf) | ISBN 9780784414583 (soft cover : alk. paper) | ISBN 9780784480236 (PDF)
Subjects: LCSH: Pipelines–Design and construction. | Trenchless construction. | Boring.
Classification: LCC TA660.P55 (print) | LCC TA660.P55 M53 2017 (ebook) | DDC 621.8/672–dc23
LC record available at https://lccn.loc.gov/2016044510

Published by American Society of Civil Engineers
1801 Alexander Bell Drive
Reston, Virginia, 20191-4382
www.asce.org/bookstore | ascelibrary.org

Errata: Errata, if any, can be found at https://doi.org/10.1061/9780784414583.

ISBN 978-0-7844-1458-3 (print)
ISBN 978-0-7844-8023-6 (PDF)

Manufactured in the United States of America.

23 22 21 20 19 18 17 1 2 3 4 5

MANUALS AND REPORTS ON ENGINEERING PRACTICE

(As developed by the ASCE Technical Procedures Committee, July 1930, and revised March 1935, February 1962, and April 1982)

A manual or report in this series consists of an orderly presentation of facts on a particular subject, supplemented by an analysis of limitations and applications of these facts. It contains information useful to the average engineer in his or her everyday work, rather than findings that may be useful only occasionally or rarely. It is not in any sense a "standard," however; nor is it so elementary or so conclusive as to provide a "rule of thumb" for nonengineers.

Furthermore, material in this series, in distinction from a paper (which expresses only one person's observations or opinions), is the work of a committee or group selected to assemble and express information on a specific topic. As often as practicable the committee is under the direction of one or more of the Technical Divisions and Councils, and the product evolved has been subjected to review by the Executive Committee of the Division or Council. As a step in the process of this review, proposed manuscripts are often brought before the members of the Technical Divisions and Councils for comment, which may serve as the basis for improvement. When published, each work shows the names of the committees by which it was compiled and indicates clearly the several processes through which it has passed in review, so that its merit may be definitely understood.

In February 1962 (and revised in April 1982), the Board of Direction voted to establish a series titled "Manuals and Reports on Engineering Practice," to include the Manuals published and authorized to date, future Manuals of Professional Practice, and Reports on Engineering Practice. All such Manual or Report material of the Society would have been refereed in a manner approved by the Board Committee on Publications and would be bound, with applicable discussion, in books similar to past Manuals. Numbering would be consecutive and would be a continuation of present Manual numbers. In some cases of joint committee reports, bypassing of Journal publications may be authorized.

A list of available Manuals of Practice can be found at http://www.asce.org/bookstore.

CONTENTS

PREFACE

There are several different trenchless technologies. Horizontal auger boring (HAB) is one of these and is the focus of this manual of practice (MOP). The first ASCE MOP for auger boring projects was developed in 2004 by the Auger Boring Task Force of the ASCE Committee on Trenchless Installation of Pipelines (TIPS). This MOP was the first in a series of MOPs that have promoted best practices and developed a knowledge base for auger boring projects. Since then, there have been advancements in HAB technology. Therefore, the Auger Boring Task Force established by TIPS has issued this updated and comprehensive edition.

This MOP will assist engineers, contractors, and owners involved in new pipe installation projects that use the HAB method to design and carry out projects effectively and safely, in conformance with project requirements and site conditions. The objective of this manual is to present a clear understanding of the method, its capabilities, and limitations; outline important design and construction considerations; and identify potential problems along with prevention and mitigation measures. The task committee understands that various trenchless technologies may be combined to form hybrid trenchless methods. However, this MOP will focus on the mechanics of the basic HAB means and methods with currently available equipment. These guidelines are based on information compiled from manufacturers' literature, field experience, technical papers, and other related information, and from comments and reviews made by the Blue Ribbon Committee.

The task committee would like to thank all the task committee members and reviewers for their support, time, and effort.

ACKNOWLEDGMENTS

Prepared by:	The Horizontal Auger Boring Task Force
Under the Direction of:	The Trenchless Installation of Pipelines (TIPS) Committee of the Utility Engineering and Surveying Institute of the American Society of Civil Engineers
Task Committee Chair/Editor:	Alan Atalah Bowling Green State University, Bowling Green, Ohio
Task Committee Secretary:	Lameck Onsarigo Kent State University, Kent, Ohio
Task Committee Members:	Craig Camp Hatch Mott MacDonald, San Diego, California Tom Fuerst The Robins Company, Solon, Ohio Robert Martin CH2M Hill, Milwaukee, Wisconsin Babs Marquis McMillen Jacobs Associates, Burlington, Massachusetts David Scarpato Scarptec, Inc., Monument Beach, Massachusetts Nick Strater Brierley Associates, Bedford, New Hampshire

REVIEW COMMITTEE

Leader:	Tom Iseley, Ph.D., P.E., Dist.M.ASCE CETF Professor, Civil Engineering Director, Trenchless Technology Center Louisiana Tech University, Ruston, Louisiana
Committee Members:	Mohammad Najafi, Ph.D., P.E. University of Texas at Arlington, Arlington, Texas Brian Dorwart, P.E., P.G. Brierley Associates, Bedford, New Hampshire Daniel L. Liotti, P.E. Midwest Mole, Inc., Greenfield, Indiana

CONTRIBUTORS TO THE ORIGINAL MANUSCRIPT (2004)

Task Committee Chairman/Editor:	Mohammad Najafi University of Texas at Arlington, Arlington, Texas (formerly with Michigan State University)
Task Committee Secretary:	Ossama Salem University of Cincinnati, Cincinnati, Ohio
Authors and Sponsors:	Jim Barbera Barbco, Inc., Canton, Ohio Leo Barbera Horizontal Holes International, Inc. Southern Pines, North Carolina Guru Kulandaivel Center for Underground Infrastructure Research and Education, Michigan State University, East Lansing, Michigan Dan Liotti Midwest Mole, Inc., Indianapolis, Indiana Michael J. Moore McLaughlin Boring Systems, Greenville, South Carolina Carl Neagoy Akkerman Inc., Brownsdale, Minnesota Gaylord Richey Astec Underground/American Augers, Loudon, Tennessee Raj Tanwani Ratnala-Bahl Engineers, Houston, Texas

ASCE COMMITTEE ON TRENCHLESS INSTALLATION OF PIPELINES (TIPS)

Chair:	Tennyson Muindi, P.E.
Staff:	Donna Dickert Acquisitions Editor Tenzing Barshee Program Coordinator, Technical Activities
Technical Editor:	Julie McCullough McMillen Jacobs Associates, San Francisco, California

ABBREVIATIONS, ACRONYMS, AND INITIALISMS

A/O TV	acoustic/optical televiewer
AASHTO	American Association of State Highway Transportation Officials
ANSI	American National Standards Institute
ASCE	American Society of Civil Engineers
ASTM	American Society for Testing and Materials
BMTA	boring machine tunnel attachment
BVR	boulder volume ratio
CBS	controlled boring system
CCFRPP	centrifugally cast fiberglass reinforced plastic pipe
CVR	cobble volume ratio
DOT	Department of Transportation
DSC	differing site condition
DTH	down-the-hole
EPA	U.S. Environmental Protection Agency
FEMA	Federal Emergency Management Agency
FHWA	Federal Highway Administration
GBM	guided boring method
GBR	geotechnical baseline report
GDM	geotechnical design memorandum
GDR	geotechnical data report
GIR	geotechnical interpretive report
GPR	ground-penetrating radar
GWT	groundwater table
HAB	horizontal auger boring
HEB	horizontal earth boring
HDD	horizontal directional drilling

HDPE	high-density polyethylene
HSA	hollow stem auger
ISRM	International Society of Rock Mechanics
ISTT	International Society for Trenchless Technology
LDBA	large-diameter boring attachment
LED	light-emitting diode
MIG	metal inert gas
MOP	manual of practice
MT	microtunneling
MTBM	microtunneling boring machine
NASTT	North American Society for Trenchless Technology
NDT	nondestructive testing
NPDES	National Pollutant Discharge Elimination System
OSHA	Occupational Safety and Health Administration
PI	plasticity index
PJ	pipe jacking
PTM	pilot tube method
PVC	polyvinyl chloride
QL	quality level
RPMs	revolutions per minute
RQD	rock-quality designation
SBU	small boring unit
SBU-A	SBU-auger
SBU-M	motorized SBU
SBU-RH	SBU rock head
SPT	standard penetration test
SUE	subsurface utility engineering
TBM	tunnel boring machine
TIPS	trenchless installation of pipelines
UCS	unconfined compressive strength
USACE	U.S. Army Corps of Engineers
USGS	U.S. Geological Survey
UT	utility tunneling
UTRC	Underground Technology Research Council

CHAPTER 1
GENERAL

There is a significant demand for underground infrastructure to transport our drinking water, wastewater, stormwater, natural gas, oil, and power. Once every four years, the American Society of Civil Engineers (ASCE) provides a comprehensive assessment of the nation's major infrastructure categories in the *Report Card for America's Infrastructure*. Using a simple A to F school report card format, the *Report Card* provides a comprehensive assessment of current infrastructure conditions and needs, assigns grades, and makes recommendations for how to raise the grades (ASCE 2013).

According to the 2013 *Report Card*, the grade for drinking water, wastewater, and stormwater systems improved slightly from a D– to a D from 2009 to 2013. Many of our drinking water, wastewater, and stormwater systems are nearing the end of their useful life. There are an estimated 240,000 water main breaks per year in the United States. If every pipe needed to be replaced, the cost over the coming decades could reach more than $1 trillion. Capital investment needs for the nation's wastewater and stormwater systems alone are estimated to total $298 billion over the next 20 years. Pipes represent the largest capital need and comprise three-quarters of the total infrastructure needs. Rehabilitating to extend life spans and upgrading and expanding the network with new pipes will address many of the sanitary sewer overflows, combined sewer overflows, and other pipe-related issues (ASCE 2013).

The grade for energy remains a D+; America relies on aging pipeline transmission and distribution systems, some of which originated in the 1880s. The increased demand for natural gas and oil will become a greater challenge after 2020 as the population increases (ASCE 2013). Even with substantial shifts to "green" energy sources, energy still requires transmission from its source and distribution to the consumer.

Invisible underground infrastructure is the foundation of our current civilization and future prosperity. They permit the concentration of resources required to grow our economy. The underground infrastructure is failing due to a lack of proper maintenance and asset management practices. Underground infrastructure has been getting less attention than aboveground infrastructure because underground infrastructure is out of sight and out of mind (Salem et al. 2008). However, it requires maintenance, repair, rehabilitation, and replacement in an economical and safe manner that causes minimal capital, environmental, or social costs to taxpayers.

Proper maintenance requires the periodic inspection of these underground assets. After inspection, remedial action is taken, as needed, and may include cleaning, repair, rehabilitation, or replacement. Significant improvements in pipe rehabilitation technologies in recent years have provided a significant cost benefit to extend pipeline life, but new installations remain a critical component of upsizing and expanding our infrastructure to handle today's rapid population growth. However, if a pipe needs to be replaced or a new pipe is needed, construction is limited to the traditional open-cut methods or the trenchless technologies. Traditional open-cut construction involves the following activities: exploring existing utilities, diverting traffic, excavating, shoring, dewatering, laying the pipe, backfilling, and restoring the site. Excavated urban materials are often contaminated and may need to be disposed of at an off-site location, requiring backfill material to be transported to the site. In urban areas, these activities can cause significant disturbance to traffic, the environment, local businesses, and residents; therefore, trenchless technology is the preferred option for pipeline installation.

The North American Society for Trenchless Technology (NASTT) defines trenchless technology as a family of construction techniques for installing or rehabilitating underground infrastructure with minimal disruption to traffic, businesses, and residents (NASTT 2013). This definition also includes technologies for inspection, leak detection, and leak location with minimal disruption and minimal excavation from the ground surface. When employed appropriately, trenchless technology is an economical and convenient alternative to open-cut construction (Atalah and Kariuki 2009).

Chapter 1 of this MOP discusses the importance of trenchless technology and its role in addressing the water, wastewater, oil, and gas pipeline issues mentioned above. The chapter presents an overview of the horizontal auger boring (HAB) process, along with alternative trenchless techniques used to install new water and wastewater lines.

1.1 SCOPE OF THIS MOP

This MOP presents HAB technology and its application to utility construction projects. Chapter 2 of this manual examines the mechanics of the HAB system, including the different components and attachments, along with recent innovations in the technology. The rest of the chapters present and discuss the different phases of a HAB project: the planning phase (Chapter 3), the geotechnical investigation process (Chapter 4), the design and preconstruction phase (Chapter 5), and the project construction phase (Chapter 6).

It is always a challenge to determine the level of detail in an MOP for a technology like HAB because it is highly adaptable and is applicable over a wide range of conditions. HAB can be used to install simple, short, and unsteered casings in uniform ground. The method can also be used to install complicated long crossings in challenging ground and environments. The level of project risk typically determines the complexity. Low-risk projects are often completed with little preconstruction phase activity (planning, geotechnical exploration, and design); the assumption is that the risks will be dealt with if and when they are encountered. Risky and challenging projects typically involve extensive planning, exploration, and design prior to the construction phase. This manual is written to address the installation of a typical carrier pipe with low to moderate risks and challenges. For these projects, the owner and engineer are typically faced with a series of judgment calls on the number of preconstruction activities needed to identify the level of risk involved, the challenges, the parties' experience, the project's contracting mechanisms, and the associated environmental and social issues.

1.2 WHY TRENCHLESS TECHNOLOGY?

More and more projects are now being assessed by processes called lifecycle cost or triple bottom line costing. These methods assess the total project cost, including the direct, premium, operation, and maintenance costs over the project life, along with the social and environmental costs. When premium, social, and environmental costs are added, trenchless technology costs become lower than open-cut technology costs. It is these total cost savings that have been driving the past and current rapid market growth of trenchless technologies.

Direct costs—which consist of the costs of labor, material, equipment, and associated business expenses—are an important factor in the selection of the construction method. The direct costs of a pipeline project depend on direct controllable factors, including the pipe size (diameter), the material, and the length and depth of the installation. Direct costs also depend on indirect controllable factors, including the project location, the project schedule, the surface and subsurface conditions, the pipeline application, and the weather conditions. These indirect factors impact cost and schedule and thus create perceived risks that need to be included in project costs. Remember, good contractors do not take on risk, they price risks.

Premium costs—which consist of increased costs above locally typical cost caused by project-specific conditions, such as work interval, schedule restrictions, and subsurface features like higher excavation difficulty and utility location and avoidance—are an important factor in the development of the construction capital cost of a project.

Operation and maintenance costs over the project life, or lifecycle costs, have been studied for many years. At issue is often the true life expectancy of a project, based on all project impacts. The assessment of these costs is typically not influenced strongly by the selection of open-cut or trenchless construction techniques.

Social and environmental costs—which consist of costs to address indirect social and environmental costs—are real costs to the taxpayer but are not included in the capital funding of a project. These costs are the ones associated with phone calls to the mayor or other public officials resulting from negative effects on the public. The social costs of the trenching operations may include traffic interruption, business disruption, noise and dust generation, negative environmental impacts, safety hazards, pavement service life reduction, and damage to existing utilities and structures. Environmental costs may be impacts on habitat and pollution and may include carbon and other emissions.

The use of trenchless technologies has the advantage of reduced disruption to traffic, the environment, existing utilities, and economic activities, as well as reduced carbon emissions. Therefore, trenchless technologies are particularly attractive in urban areas with heavy vehicular and pedestrian traffic, small businesses, and numerous existing underground utilities. Trenchless construction is an attractive option for crossing transportation corridors (roadways, railways, and waterways). Trenchless techniques can be used to install, rehabilitate, or replace utilities located in environmentally sensitive areas and in locations where surface access may be restricted because of existing structures, environmental constraints, or landscape. Often, at these types of project sites, trenchless techniques are the only viable construction option. Trenchless techniques are often the least disruptive as well as the least costly options for projects in congested and environmentally sensitive areas (ISTT 2013c). In this light, trenchless technologies are gaining popularity in utility construction.

1.3 HORIZONTAL AUGER BORING

The International Society for Trenchless Technology (ISTT) defines HAB as a technique for forming a bore from a start or drive shaft to a reception shaft by means of a rotating cutterhead. Spoils are transferred from the face to the drive shaft by helically wound auger flights rotating in a steel casing (ISTT 2013a, b). The basic components of an auger boring system are the base unit, the casing pusher, the power pack, the auger sections, the track, and the track extensions. Different cutterheads are available for different subsurface conditions.

The HAB method is generally used to install steel casing pipe in relatively soft and stable soil conditions, such as clay, silt, and sand. Today's cutterhead technologies have also made it possible to install casings in some rock conditions above the water table. During operation, the HAB machine advances along a track that is aligned to drive the casing pipe along a designed path, as shown in Fig. 1-1.

The HAB installation cycle is based on the machine excavating and advancing the cutterhead until it reaches the designed end. Once the machine pushes the casing to the end of the track, the auger flight is disconnected from the machine and the machine is retracted, allowing a new auger flight to be connected to the existing flight and a new casing segment to be welded to the existing casing pipe. The auger string and the casing are then connected to the HAB machine, and the excavation and thrust process is repeated until the installation is completed (ISTT 2013a). The auger chain is then withdrawn from the casing pipe, and the casing pipe is cleaned of the spoils. At this time, there is an option to leave the casing pipe in the ground or to replace the casing pipe with a different pipe and pipe material by placing the new pipe on the end of the existing casing

Fig. 1-1. Layout of a HAB operation

and pushing it through the constructed bore while removing the steel casing from the receiving pit. The carrier pipe is then installed, and the annular space between the casing and the carrier pipe is grouted or filled with granular material (if desired). There are two main methods for HAB: the cradle method and the track method.

The cradle method does not require any operations to be performed in the pit. Instead, the pit is excavated to contain the spoils, and a pipeline side boom typically holds the cradle auger boring machine. This method requires a large launch pit, which spans the full length of the casing pipe necessary for the crossing and collects the spoils. Thrust for the HAB machine is provided from a hydraulic winch on the machine, and the cable is hooked to a deadman anchor located at the bore entrance, as shown in Fig. 1-2. The HAB machine spins a full-length auger contained in the casing, which moves the cuttings to the discharge at the HAB end of the casing. Cradle auger boring is used mostly to install transmission lines under roads, railroads, and so forth and can be used for bores up to about 200–250 ft (60–76 m) in length.

The track method involves the installation of a HAB machine in the jacking pit on a predetermined length of track. Track lengths are typically based on the length of the machine plus the design pipe section length for the project. The HAB machine consists of a power pack, a master casing pusher, a base push unit, and a saddle. The jacking force is provided by the hydraulic thrust, as shown in Fig. 1-3, reacting against a thrust block or base slab that supports the track system. The casing pipe and auger sections are added as the machine advances, and the spoils are removed

Fig. 1-2. Cradle auger boring machine in operation

Fig. 1-3. Layout of a track-type HAB operation
Source: Osborn Engineered Products (2016)

through the use of augers inside the casing and transported to the entry pit, where the spoils are ejected via a door on the side of the machine.

1.3.1 Auger Boring Process

The auger boring process begins with advancing the lead casing, with the cutterhead at the front of the casing. As the cutterhead cuts through the soil, it directs the soil toward the augers inside the casing. The augers transport the excavated material through the casing from the back end of the cutterhead to the receiving shaft, where the material is expelled via spoils ejector paddles through a trap door on the master casing pusher. When the lead casing has been advanced into the soil to the limit of the track, another casing (with the augers inserted in it) is placed on the track. The augers inside the new pipe are connected to the ones inside the lead casing via pins, as shown

Fig. 1-4. Connecting auger sections

in Fig. 1-4. The two pipes are welded, and the auger boring proceeds by jacking the casings while simultaneously rotating the cutterhead and transporting the spoils to the jacking pit via the encased augers.

The ground condition dictates the advancement speed of the HAB machine. An experienced HAB machine operator matches the rate of spoils removal with the rate of HAB machine advancement to optimize the production rate. The operator, through experience, monitors the ease or difficulty with which the cutterhead cuts the soil, along with the amount of ejected spoils. On the basis of this information, the operator adjusts the excavation to advance efficiently without over- or underexcavating, which can result in settlement or heave.

The HAB machine operator uses a water-level system, laser level, or electronic guidance system to monitor the machine's vertical location. Steering is achieved by using manual shoes that are adjusted by personnel entry into the casing or, if an articulated head is installed, by using mechanical rods and hinges at the articulated front end of the HAB machine to steer the machine by deflecting the articulated head up and down. In longer drives, a lubricant can be injected into the annular space between the excavated ground and the outside surface of the casing to reduce friction and jacking forces. See Chapter 6 for more discussion on the steering, guidance, and lubrication systems.

These processes, cutting the soil, transporting the spoils, jacking the casing, and welding the new casing segment, are repeated until the cutterhead reaches the designated receiving shaft. Then, the cutterhead is disconnected and removed, the augers are removed from the casing (bringing along the remaining spoils to the jacking shaft), and the pipe is cleaned. If the casing is intended to be the carrier pipe, the pipe is ready to be used. Otherwise, a carrier pipe is inserted into the casing and adjusted for the designed grade; then, the annular space between the casing and the carrier pipe may be filled with sand or grout or may remain unfilled, and both ends of the casing are sealed with a bulkhead that may be vented to the ground surface.

1.3.2 HAB History

The HAB method is used extensively throughout the United States. Although this method has evolved over the last half century, the major developments have taken place over the last two decades. These developments have led to increased bore diameters, lengths, accuracy, and ability to bore through different conditions.

Like most other construction equipment, the HAB machine was developed because of market demand. As early as 1936, the first boring machines were made from the rear ends of old trucks. CRC-Evans Pipeline International was the first to build these machines, which were cradle auger boring machines, similar to the one shown in Fig. 1-5. These HAB machines were suspended by pipe-laying equipment and required a jacking lug (deadman) in the ground in front of the machine. Cables tied to the jacking lug were used to provide the thrust needed to pull the machine toward the borehole.

The HAB method was formally developed in the 1940s by Vin Carthy, Salem Tool Company, and Charlie Kandal. These developments were simultaneous and independent of one another. The machines were originally developed to drill horizontal blast holes into open-cast coal mines in Somerset, Pennsylvania, that contained a thick seam of high-grade coal.

In the 1940s, Charlie Kandal founded the Ka-Mo company, which manufactured the Ka-Mo HAB machines. The initial machines were

Fig. 1-5. Cradle HAB machine

powered with electricity, but later models were gasoline powered. These HAB machines were large and equipped with a drop transmission that had three speeds for advance and one speed for reverse. The Ka-Mo machine included an electric or air motor and was designed to run on a track. A bar, called a "hand croud," was used to pry the machine forward. The bores at that time were normally not cased. Often, the machine had to be held back to prevent it from screwing itself into the ground. Larger machines with approximately 16-hp air motors were equipped with two hand crouds. By 1951, Ka-Mo had established the capability to bore 5-in.-diameter (130-mm) uncased holes of up to 230 linear feet (70 m).

Salem Tool Company and Ka-Mo dominated the horizontal earth boring (HEB) business until 1961, when pipeline welder Al Richmond also started manufacturing small HEB machines. Richmond was working for a company called the Wert & Starn Pipeline Company, which was building boring machines under the trade name of Tornado. The Tornado was built with the engine on top of the gearbox and was driven with a chain. The machine had metal plates and slid in a track or on the ground. The process required a deadman in the ground in front of the boring machine, similar to the CRC machines.

Richmond built his own machines, which were similar to those of Ka-Mo. These machines also ran on tracks and were equipped with a hand croud. The tracks were built with holes in the cross members, and T-pins were used to hold the track in the ground. He later changed the configuration of the machines and built bigger machines with hydraulic cylinders on them. The Richmond machines were the first centerline-driven machines—engine, transmission, gearbox, and hydraulic motor in one straight line—and had an automotive-type clutch instead of an over-center clutch. Richmond's main competition at that time was Ka-Mo, which was still building large machines for small-sized installations. Ka-Mo was also building its own transmissions and gearboxes; Richmond was using standard off-the-shelf items, which were less expensive and easier to replace. This difference made the Richmond small units, including the 24-in. (610-mm) Power Midget, more convenient and increasingly popular. Richmond also built 30- and 36-in. (760- and 915-mm) machines for larger diameter installations.

Since a significant amount of uncased boring was being done at that time, all machines had one hydraulic cylinder. As cased bores became commonplace, the thrust requirements increased, and Richmond used two cylinders to operate the machines. These cylinders were placed on top of the tracks. Richmond later modified the design so that the cylinders were below the track and later, very reluctantly, split the machines for ease of handling. Until 1967, Richmond machines used Ka-Mo augers and cutterheads. He then built his own augers and cutterheads, which were the same as those of Ka-Mo. Richmond continued making advances in equipment design to keep pace with the changing market.

Roy Woodruff and Son designed the split machine and put a spacer in the 36-in. machine to make it a 48-in. machine. In the 1960s, Ernie Coppica, Wixom, and Michigan invented a steering system.

In 1971, Leo Barbera left Richmond Manufacturing and founded American Augers. American Augers started by building hydrostatic drive machines with a built-in slip clutch. If the machine hit a boulder or some other obstruction, the machine would switch to maximum torque but at the same time block out the thrust. The machine would advance only when the condition was relieved.

Opportunity is the mother of invention. This has been especially true in the auger boring business, in which equipment versatility and innovative contractors have adapted this simple machine to provide not only a wide variation in subsurface conditions but also a wide variation of sizes at a low contractor investment cost. In 1995, the guided boring method (GBM), also known as the pilot tube method (PTM), was introduced in the United States from Europe and was used in conjunction with the HAB machine to install small-diameter pipes with greater line and grade precision (Martin and Grolewski 2011). The small boring unit (SBU) was developed by the Robbins Company in 1996, during the excavation of a 36-in.-diameter (915-mm) tunnel in hard rock as part of the work on the Pennsylvania Turnpike. Crews from the John Fithian Contracting Company of Youngstown, Ohio, encountered rock that proved too tough for the carbide-bit Christmas tree head on their HAB machine. Facing an impasse, Fithian brought the problem to Robbins, and the device that would ultimately become the SBU completed the drive in less than 1 week (Veidmark and Sivesin 2009).

Between 1996 and 2010, bigger and more powerful HAB machines were designed to accommodate up to 96-in.-diameter (2,440-mm) crossings. Improvements were made to enhance tool wear. Steering improvements that upgraded accuracy and performance were also made (Martin and Grolewski 2011). The industry has made strides in improving the efficiency of the engines and has introduced automatic transmission, which reduces fuel consumption and carbon footprint. Other manufacturers of HAB machines include McLaughlin Boring Systems, Horizontal Equipment Manufacturing Inc., Michael Byrnes Manufacturing, Bor-It Manufacturing, Barbco, and several other smaller manufacturing firms. Several contractors also manufacture their own HAB equipment.

1.3.3 Applications and Limitations of HAB

HAB provides an adaptable system that is often the most economical solution when the carrier pipe needs to be encased in concrete under railways and roadways or any other obstructions. HAB offers sufficient accuracy to install gravity carrier pipes inside the steel casing. The technique offers redundant protection for pressure line applications, especially when high-pressure lines cross public areas that warrant additional protection. HAB can be used today in various ground conditions because of recent innovations in cutterhead technology that have extended the capabilities of the HAB machine to penetrate medium to hard rock conditions and some mixed ground conditions.

Subsurface materials suitable for the HAB process range from dry sand to firm clay to solid rock; however, firm silty sandy clay is optimal. The rule of thumb for cobbles and boulders is that the method can accommodate cobbles and boulders up to about 33% of the casing diameter, though nested cobbles and boulders can be quite costly to excavate and can lead to adverse settlements. Even though cobbles and boulders may be able to pass through the auger, these obstructions can adversely affect the alignment of the casing. Cobbles and boulders larger than 33% of the casing diameter may stall and halt the boring process and require rescue intervention, such as surface excavation and the removal of the obstruction or the removal of the auger system and advancement by hand mining and hand splitting of the obstruction. This option requires a cutterhead that can be removed from the casing to permit the obstruction to be hand mined.

Generally, HAB is not compatible with cohesionless soils below the groundwater elevation but has marginal success with groundwater up to about one-third above the casing invert. However, granular layers interbedded with cohesive soils can potentially be mined with this technique by pulling the cutterhead back into the casing about three diameters to form a soil plug in the leading section of the auger.

HAB machines that use tungsten carbide teeth, such as those on a Christmas tree cutterhead, can cut soft rock up to 4,000 psi (28 MPa) fairly efficiently (Long 2006). Recent developments include cutterheads with disc cutters that enable HAB machines to excavate rock with an unconfined compressive strength (UCS) of 25,000 psi (170 MPa) and greater (Foster 2009).

Although the HAB method was initially developed to cross under a two-lane paved roadway with an average length of 40 ft (12 m) and a maximum length of 70 ft (21 m), several innovations have greatly enhanced the method's capability; these innovations include hydrostatic drives, lubrication injection, and higher hydraulic thrust capacity. The average bore length can range between 175 and 225 ft (53 and 68 m), with the maximum bore span greater than 600 ft (180 m).

The accuracy of the bore largely depends on operator skill, the selected steering and tracking system, the frequency of the acquiring steering position, care in the initial setup, and ground conditions with minimal obstructions and ground stiffness variance. A line accuracy of ±1% of the length of the bore is normally achieved in grade-monitored and steered HAB projects. However, grade accuracy can be increased by using grade steering systems. Line accuracy is typically achieved through the careful setup of the machine to a project azimuth. Articulated right–left steering shoes that can be adjusted for grade are also available. To properly steer right and left, it is necessary to know the location of the cutterhead relative to the project azimuth, often accomplished through the use of a horizontal directional drilling sonde that is attached to the head of the casing. A walk-over readout located on the ground surface can then determine the horizontal position and depth below the ground surface of the sonde. Alternatively, an oversized casing could be installed in situations in which a higher level of accuracy is required. The oversized casing provides some extra room to adjust the carrier pipe within the casing by blocking to achieve the required carrier pipe accuracy. Recent innovations in the guidance systems have made it possible to achieve greater levels of both line and grade accuracy. However, these technologies also involve greater costs (Najafi 2013).

1.4 OVERVIEW OF ALTERNATIVE TRENCHLESS TECHNOLOGIES

Several construction techniques and methods qualify as trenchless (ISTT 2013c). They can be broken down into two basic groups: (1) replacement and rehabilitation and (2) new installations. Replacement and rehabilitation techniques require an existing pipe in the ground; new installation techniques do not. The replacement techniques include pipe bursting, pipe splitting, slip lining, and pipe reaming. Rehabilitation techniques include cured-in-place pipe, deformed (fold and form) pipe, spiral wound pipe, hand or pneumatically applied geopolymer and portland products, composite epoxy liners, and segmental linings.

As shown in Fig. 1-6, the new installation methods can be divided into two groups on the basis of the steering capabilities of each method. The nonsteerable systems include pipe ramming, piercing tools, and some forms of HAB, both with and without casing. The steerable methods include horizontal directional drilling (HDD), pipe jacking (PJ), microtunneling (MT), utility tunneling (UT), the pilot tube method (PTM), and HAB. The steering ability in that context may not be sufficient for gravity installations, which require a maximum deviation of 1 in. (25.4 mm) on line and grade.

Each trenchless technique has capabilities and limitations. The selection of a trenchless method for a particular application depends on several factors, such as

- Type of proposed utility,
- Pipe material requirements,

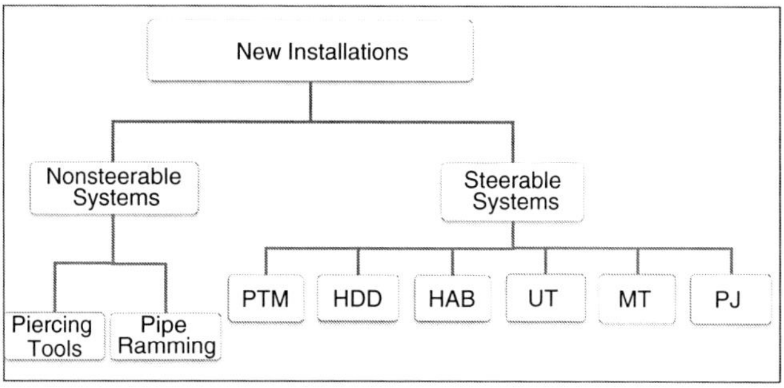

Fig. 1-6. Major divisions of trenchless techniques
Source: Atalah (1996)

- Groundwater table (GWT),
- Available work space (horizontal and vertical),
- Required line and/or grade accuracy,
- Diameter of utility,
- Environmental constraints,
- Subsurface conditions,
- Existence and status of old pipe and other utilities in the immediate vicinity,
- Depth of pipeline,
- Dewatering requirements, and
- Availability of experienced local contractors (available trenchless system, equipment size, team experience, and availability of pipes and accessories).

As a comparison to HAB, the other pipe installation methods—namely, microtunneling, pipe jacking, HDD, pipe ramming, utility tunneling, and pilot tube—are briefly discussed below. The microtunneling section includes more detail than is given for the other systems. These details are common to the subsequent systems and are thus presented in the microtunneling section, then referred to in the subsequent sections.

1.4.1 Microtunneling

ASCE's *Standard design and construction guidelines for microtunneling* (ASCE 2015) defines microtunneling as a trenchless construction method for installing pipelines that meets the following criteria:

- The microtunneling operator remotely operates the microtunneling boring machine (MTBM) from a control panel, normally located on the surface. Therefore, personnel entry into the pipeline is not required for routine operation.

- The microtunneling system utilizes a guidance system, usually consisting of a laser beam projected onto a target on the back end of the MTBM, enabling the system to install gravity sewers or other types of pipelines to the required tolerance for line and grade.
- The pipeline is constructed by consecutively pushing pipes and the MTBM through the ground through the use of a jacking system for thrust.
- Continuous pressure is provided to the face of the excavation to balance groundwater and earth pressures (ASCE 2015).

There are two types of microtunneling methods:

- Slurry-type MTBM: The spoils are mixed with slurry in a chamber located behind the cutterwheel, and are conveyed back to the ground surface. The pressurized slurry balances the earth and the groundwater pressure to support the excavation face and thus limit settlement.
- Auger-type MTBM: The spoils are conveyed from the MTBM to the jacking shaft by a continuous-flight auger enclosed in a separate casing inside the pipe being jacked. Positive pressure is maintained at the excavated face to support the excavation face by regulating the volume of removed spoils relative to the advance rate of the MTBM.

Slurry microtunneling systems are used in installations below the GWT in granular soils, and auger microtunneling systems are usually used above the GWT, slightly below the GWT, or in cohesive soils with low permeability. Microtunneling may also be completed in bedrock, provided that the machine and cutterhead used are designed for excavating the rock and that the slurry or auger system is designed to remove the resulting spoils.

Typical microtunneling installations range in length from about 200 to 500 ft (61 to 152 m); longer installations are achievable, depending on the machine capabilities, diameter, intermediate jacking stations, and jacking force requirements. Although curved microtunnels are possible, they are technically challenging with a customized pipe design and relatively expensive. The vast majority of microtunneling installations being completed in North America at the time of this publication are straight.

1.4.1.1 Method The main microtunneling system components are provided below and illustrated in Fig. 1-7:

- MTBM
- Transportation/spoils removal system (slurry or auger)
- Hydraulic jacks
- Guidance and remote control system
- Jacking pipe

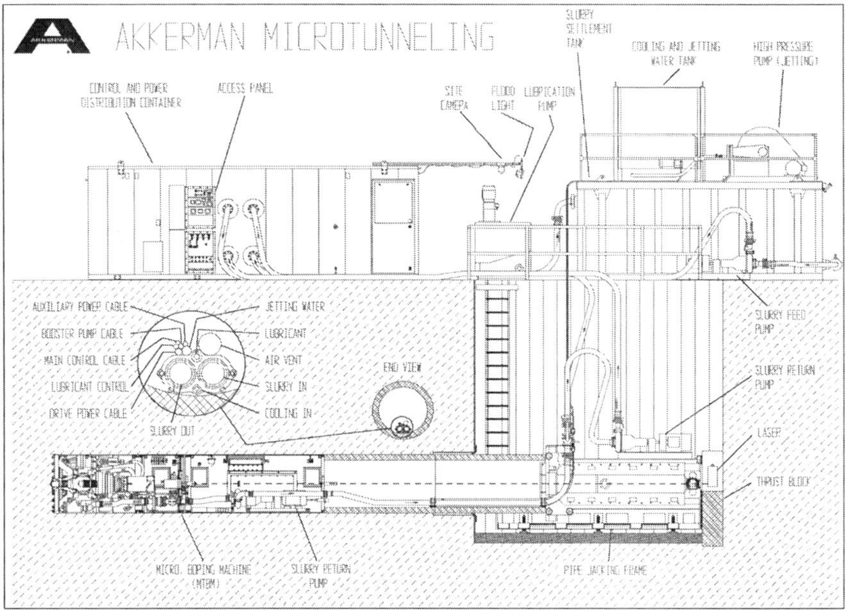

Fig. 1-7. Slurry-type microtunneling operation
Source: J. Holden, personal communication, 2016

In microtunneling, a rotating cutterhead excavates the soil mechanically and transports it to the crushing chamber (located directly behind the cutterhead), where the excavated material is crushed to less than 1 in. in size. Then the excavated material is extruded through small ports located at the rear of the crushing chamber. In the slurry-type method, the excavated material is mixed with a pressurized slurry, transported through the pipes (placed inside the jacking pipe) to the jacking shaft, and then pumped to the spoils separation unit. The spoils are removed from the slurry and then recycled, or new slurry is transported to the face of the excavation. In the auger-type method, the excavated material is transported to the jacking shaft by augers in a cased pipe (placed inside the jacking pipe) and then hoisted to the surface by a crane (Najafi 2013).

To avoid settlement and/or heave, hydrostatic slurry pressure counterbalances the water and/or soil pressure at the face of excavation. The counterbalancing pressure has to be more than the active soil earth pressure to avoid settlement and less than the passive soil earth pressure to avoid heave. The rate of excavation (controlled by the size of the opening in the MTBM) is balanced against the rate of jacking (Atalah 1996).

The jacking system (consisting of the jacking frame and hydraulic jacks) is used to advance the MTBM and the pipe string during the boring process. It is critical that the operator control the face pressure first to avoid settlement or heave then match the advance rate to the excavation rate to optimize the production rate. A laser beam is transmitted from the jacking shaft to a target mounted at the back of the MTBM. The image of the laser on the target plate is transmitted to the operator, who steers the MTBM appropriately to keep it on the desired line and grade. With an experienced operator, an accuracy of ±1 in. is achievable (Atalah 1996).

The jacking pipe is designed to withstand predicted jacking and long-term operational forces. Maximum loads are typically experienced during construction; thus, construction and long-term loading conditions both need to be considered in the pipe design. The pipe is also designed to have smooth outside profiles and watertight joints. The material of the commonly used jacking pipes includes reinforced concrete pipe (RCP), steel pipe, ductile iron pipe, centrifugally cast fiberglass reinforced plastic pipe (CCFRPP), vitrified clay pipe (VCP), and polymer concrete pipe (PCP) (Atalah 1996).

For a typical work cycle, the MTBM is jacked into the ground simultaneously with the excavation of soil at the face. Then, the hydraulic hoses and slurry lines are disconnected or the auger discharge is moved, the jacks are retracted, a pipe segment is lowered and connected, and the slurry lines and hydraulic hoses or the auger system are reconnected. Then, excavation, jacking, and spoils removal are resumed. This process is repeated until the drive is completed and the MTBM can be removed from the receiving shaft (Nada Pacific Corporation 2013).

The typical range in diameter for microtunneling operations is 12–126 in. (305–3,200 mm); the most common range is 24–60 in. (610–1,525 mm). The method is typically used for the installation of gravity lines when a high degree of accuracy is required or in low-strength or saturated soils that become unstable when subjected to open excavation that creates an unsupported excavation face. Slurry microtunneling is the method of choice for deep gravity-based installations below the water table. Auger microtunneling is often selected for installation in low-permeability soils below the water table or granular soil above the water table when control of surface settlement is a project issue.

1.4.1.2 Advantages and Disadvantages Microtunneling has several advantages; chief among them is the ability to install the product pipes within the accuracy range required for gravity installations without the need for a carrier pipe and grouting, below or above the GWT. Microtunneling is a safer installation technique than pipe jacking or tunneling, since no personnel are needed inside the pipe during pipe installation. Most MTBMs can fail when unplanned obstructions are encountered, as most MTBMs do not allow access to the face of the excavation to remove obstructions. Cobbles and boulders that may be encountered during the drive are typically considered to be obstructions. As noted by Najafi (2013), MTBM systems have difficulties in soils with cobbles and boulders that are larger than 33% of the diameter of the machine, in locations where there are nested cobbles and boulders, or when the cobbles and boulders are not tightly restrained within a soil matrix. Some other disadvantages are the need for (1) a high capital investment to purchase and mobilize an MTBM system, (2) a skillful operator and crew, and (3) shaft and surface excavations for lateral connections (Stein 2005).

1.4.2 Pipe Jacking

ASCE defines pipe jacking as "a system that uses hydraulic jacking from a drive shaft to directly install pipes behind a shield machine so that they form a continuous string in the ground" (ASCE 2015). Pipe jacking is a trenchless technique commonly used to install gravity underground utilities, especially sewers and storm drains underneath highways, waterways, and railways.

1.4.2.1 Method As shown in Fig. 1-8, pipe jacking involves pushing pipes through the ground from an entry/launch/jacking shaft toward an exit/receiving shaft, while simultaneously excavating at the face inside a jacking shield or a tunnel boring machine (TBM). The rate of excavation must be compatible with the rate of jacking to avoid heaving or subsidence at the ground surface. However, a highly skilled excavator crew is needed to determine the shape and sequence of excavation to maintain a stable

Fig. 1-8. Manned pipe jacking operation
Source: Pipe Jacking Association (2015)

face. If necessary, there are various techniques to provide temporary face stability during the excavation process.

The outside diameter of the jacking shield is usually 0.5–1 in. (12.7–25.4 mm) larger (overcut) than that of the jacking pipe to reduce the friction between the soil and the jacked pipe. Also to reduce friction during jacking, a lubricant consisting of bentonite and/or polymers is often injected into the annular space between the jacked pipe and the surrounding soil. There are various methods of excavating and transporting the soil from the excavation face to the jacking pit, where the spoils are lifted by a crane or backhoe to be transported to a disposal site (Atalah et al. 1994).

In most cases, construction begins with the entry and exit shafts. The floor of the jacking shaft and the thrust block are designed to withstand the predicted jacking forces that will be exerted during the pipe jacking process. Reaction systems should be conservatively designed to address the risk of the need for higher jacking loads and to prevent rotation within the system that can result in misaligned forces and cause elevated jacking forces and an elevated risk of overstressing the jacking pipe. The jacking equipment, which consists of the jacking frame and the hydraulic jacks, is then installed in the jacking shaft, and the laser guidance system is installed. After the jacking shield is lowered into place, it is advanced with the jacks as the face is excavated, and the spoils are transported from the face. Excavation may be achieved by using a roadheader or a modified hydraulic backhoe (excavator shield) or by hand mining with jackhammers or picks and shovels. Spoils removal may be achieved with muck carts, a conveyor belt, augers, or a vacuum extraction system. Once fully advanced, the jacks are retracted and a new pipe segment is placed on the tracks. The jacking, excavation, and spoils removal processes are resumed. This cycle is repeated until the shield reaches the exit pit, where the shield is extracted (Atalah et al. 1994). When ground conditions are relatively uniform, the shield and excavation system may be mechanized with the use of a rotating cutterhead.

1.4.2.2 Advantages and Disadvantages Pipe jacking has several advantages, including the ability to install the product pipes within 1 in. on line and grade (the accuracy required for gravity installation) without the need for a casing and grouting. It is also an economical trenchless technique for installing pipes ranging from 42 to 90 in. (1,065 to 2,285 mm) in diameter. Larger diameter pipes and rectangular boxes can also be jacked at an economical cost. Pipe jacking allows access to the face of the excavation to remove boulders or cobbles and therefore minimizes the need for a rescue shaft. However, personnel access inside the jacking shield is required to operate the equipment and facilitate excavation at the face, and dewatering is needed in granular soils below the GWT (Stein 2005).

1.4.3 Horizontal Directional Drilling

HDD is used to create a slurry excavated bore for the installation of pipes, conduits, cables, or any other linear utility. The process uses a drilling rig located at the ground surface with an angled mast to permit the drill to advance sideways. The HDD installation may consist of up to three stages: (1) pilot bore, (2) single or multiple reaming/hole enlargement, and (3) pullback. HDD planning usually starts with determining the bore plan and profile, which generally consists of three sections: an entry sloped tangent section with a vertical curve, a flat invert tangent section, and an exiting vertical curve and sloped tangent section. The depth and entry angle along with the designed product depth determine the setback distance from the production portion or usable section of a bore path at the connection point to an actual point of entry/exit at the ground surface.

1.4.3.1 Method A typical HDD project requires rig setup in an appropriately sized work space. Once the operator determines the rig location and a constructable drill path, the drill head (with drill bit and internal tracking electronics) is connected to the drill pipe. The drill mast and drill rod are then aligned with the design drill path, and the angle of the drill pipe is set at the design angle. The drill head is then advanced into the ground by the direct thrust or the thrust and torque provided by the drill rig. The drill head location is periodically determined by electronics in the drill head and on the ground surface. This process of thrusting, rotating, locating, and steering along the target bore path is repeated many times until the drill head exits the ground, as shown in Fig. 1-9.

The drill head has two ground-cutting mechanisms: the mechanical cutterhead and the hydraulic nozzles that eject drilling fluids at the front end of the drill head. As in any tunneling process, the excavation rate and spoil removal rate must match, or the production rate will become inefficient or may stop. Behind the drill bit (within the drill head) is a tracking assembly that provides the operator with the location and angle of the drill head, which is compared with the planned location and angle. If the drill head has deviated or is about to deviate from the planned path, the operator uses the orientation of a slanted drill bit to steer the drill head, assisted by the fluid jets, as shown in Fig. 1-10.

Drilling fluid is pumped through the drill pipes to the drill head assembly, where the fluid discharges and flows back to the drill rig in the cut annulus between the drill pipes and the formation. The drill fluid has many tasks, including assisting in cutting the soil, transporting the spoils to the surface, cooling the tracking tools, cooling and cleaning the drill bit, lubricating the drill hole, and supporting the formation.

If the pilot bore is to be enlarged, a reamer (equipped with drill fluid discharging nozzles, called "jets") is attached to the drilling pipe, as shown

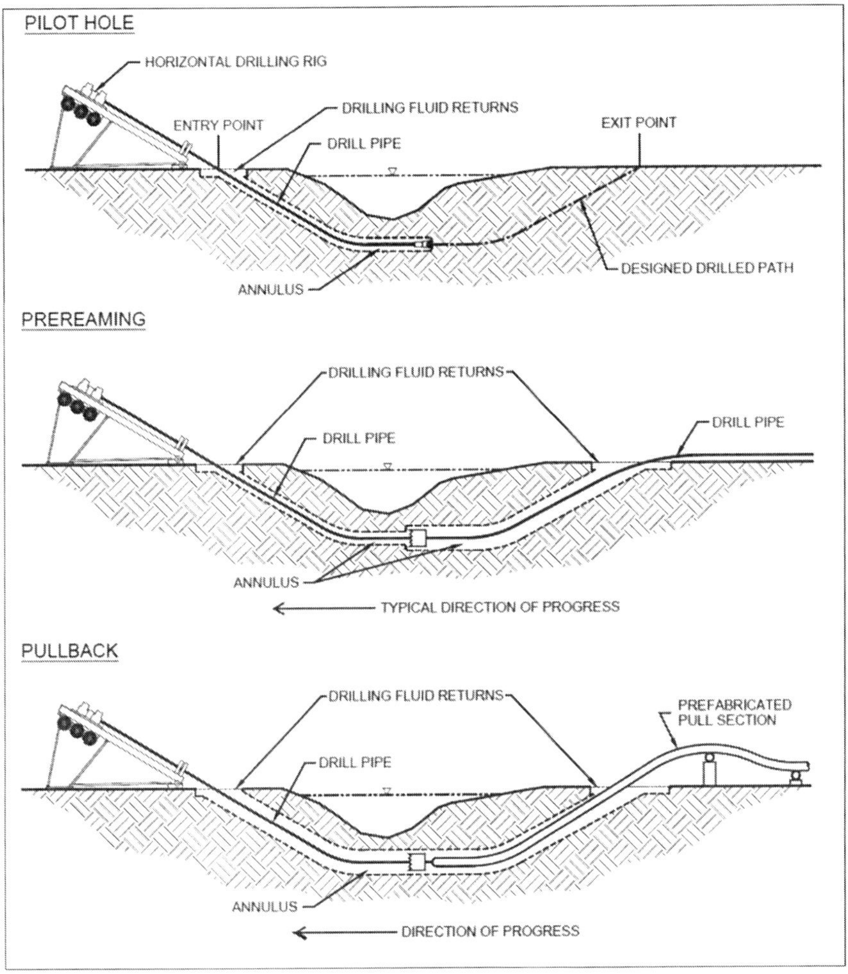

Fig. 1-9. HDD process
Source: U.S. Fish and Wildlife Service Midwest Region (2013)

in Fig. 1-11. The reamer or hole opener is rotated and pulled or pushed to enlarge the bore. The bore can be enlarged by reaming more than once, if needed, to reach the required reamed diameter. The reamed diameter is usually planned to be 1.5 times the size of the product pipe for pipes up to 24 in. (610 mm) in diameter, and 12 in. (305 mm) larger than the product pipe for larger pipes. However, experience plays a key role in sizing the larger bores to ensure successful installation of the carrier pipe. Once the desired bore size is achieved, a pulling head is attached to the preassembled carrier pipe. A swivel is placed between the carrier pipe and the

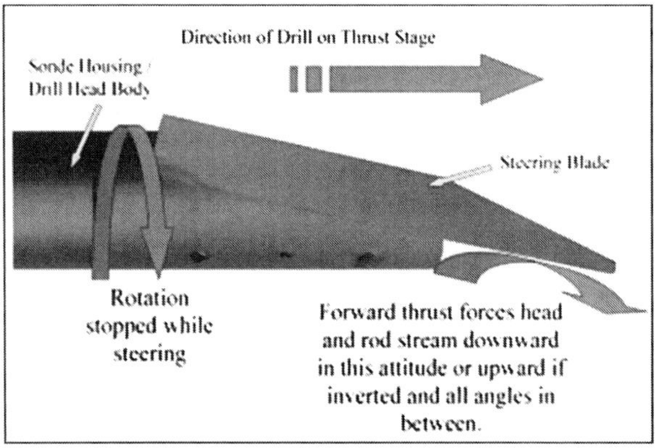

Fig. 1-10. Slanted drill bit
Source: TerraMacs (2015)

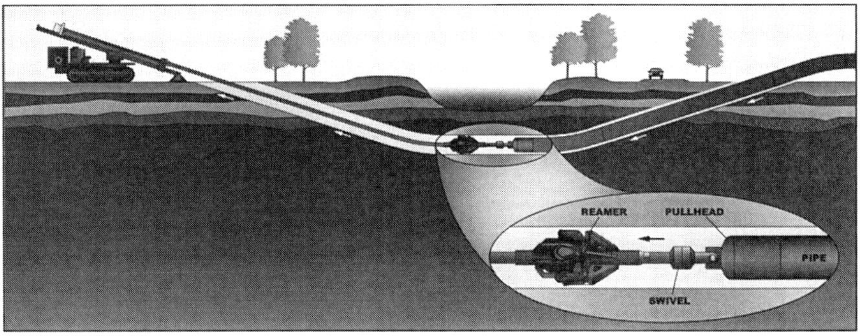

Fig. 1-11. Reaming and pullback process
Source: Mears Groups Inc. (2016)

reamer to permit the reamer to rotate without rotating the carrier pipe. This assembly is connected to the drill rig, which pulls the tools and the carrier pipe into the prepared bore. The carrier pipe must be designed to withstand the pullback forces associated with installation and potential long-term ground and operation loads. The actual force applied to the carrier pipe during installation may be monitored by load cells or by instruments on the drill rig. Load monitoring is important to make sure that the structural capacity of the carrier pipe is not exceeded, which would result in a failed installation (Bennett and Ariaratnam 2008; ASCE 2005). In the past, to protect the carrier pipe, breakaway couplers that would fail at a predetermined load were placed in the pull assembly. This process is seldom used anymore as a failure often requires the complete abandonment of the failed hole when the carrier pipe cannot be removed. Instead, if a pipe fails during installation, it may be pulled through the hole and eventually removed, allowing the hole to be recovered and the carrier pipe to be reinstalled.

1.4.3.2 Advantages and Disadvantages HDD can be an economical method for the installation of product pipes ranging from 2 to 60 in. (50 to 1,525 mm) in size in various ground conditions. Single rig installations exceeding 6,000 ft (1,829 m) in length, and double rig "intersect" bores exceeding 10,000 ft (3,048 m) in length have been achieved. Pipe materials commonly installed by HDD include high-density polyethylene (HDPE), polyvinyl chloride (PVC), ductile iron, fiberglass, and steel. In all cases, the HDD method is typically used to install utility lines and oil and gas pipelines, and to force mains below waterways, roadways, runways, and railways (ASCE 2005). Although HDD is a steerable method, it is generally not accurate enough for the installation of gravity-based utilities.

1.4.4 Pipe Ramming

Pipe ramming is generally considered to be a nonsteerable system that uses a percussive hammer to drive a steel casing from a drive pit (Read 2004). The casing may be either closed end for small-diameter pipes, usually 4–8 in. (102–205 mm) or open ended for larger diameter pipes. Casings up to 138 and 147 in. (3,505 and 3,735 mm) in diameter have been installed using this method. Typical installation lengths range from 50 to 200 ft (15 to 61 m). Pipe ramming is a common means of casing installation below railroad and highway embankments. The steel casing selected must be capable of withstanding the forces involved with the installation. Alternative pipe materials are not suitable for pipe ramming. Recently, pilot holes have been installed along a pipe ram design path and tension elements have been installed and attached to the front of the pipe to be rammed to assist in guidance and improve the accuracy of the pipe ram.

1.4.4.1 Method Installation begins with preparation of the launch pit and setting up the casing support system, or "cradle." The support system can be I-beams, wooden rail systems, or any other support system used to secure the pipe and prevent any vertical or horizontal drifting during the ramming process. The subgrade of the launching pit and the supporting system must be constructed along the target line and grade.

A cutting shoe (reinforcement steel band) is typically welded to the front end of the lead pipe to protect the cutting edge against deformation should it strike a hard object. The cutting shoe provides 0.5–1 in. (12.7–25.4 mm) of overcut, which reduces friction between the pipe and the soil. After the cutting shoe is welded to the pipe, the lead pipe is lowered onto the tracks. After verification of the gradient and the direction of the pipe and track,

Fig. 1-12. Pushing collets

Fig. 1-13. Pipe ramming operation
Source: Orton (2010), © NASTT

pushing collets (shown in Fig. 1-12), are inserted on the driving end of the pipe to lock the ramming tool/hammer into the casing, distribute the ramming force evenly to the casing, and protect the casing from damage by hammering. The tool is mechanically locked into the collets, and the hammer is activated to drive the casing into the soil. Once the casing has been fully driven, the hammer and collets are removed, the next pipe section is welded or connected, and the process is repeated as shown in Fig. 1-13 (Read 2004).

When the installation is complete, the spoils are removed and the casing is cleaned. In open-ended/open-face installation, the soil may be removed from the casing by applying compressed air or water from either end for small-diameter casings. For large casings, augers can be used to remove the soil mechanically from the inside of the pipe. For larger diameters, a small backhoe can be used (Najafi 2013). When acting as a casing for a carrier pipe, the product pipe can be inserted into the casing and the annular space can be grouted.

1.4.4.2 Advantages and Disadvantages The pipe ramming method is typically used to install steel casings under roadways, railways, and other structures. New utilities can be installed in the casing. In some instances, the method is used to install steel product lines directly. It is suitable in challenging soft soils or in rock and cobble formations smaller than the diameter of the casing. However, the method is nonsteerable; therefore, there is minimal control over line and grade, making the method unsuitable when grade is critical. Pipe ramming is risky in hard ground conditions and bedrock. In marshy soil, the steel pipe is likely to dive because of vibration-induced soil liquefaction. The pipe ramming method is loud and generates ground vibrations, which may not be suitable for adjacent vibration-sensitive structures or environments or in noise-sensitive urban areas.

1.4.5 Utility Tunneling

Utility tunneling is a general approach for constructing underground utility passages by removing the excavated soil from the front cutting face and installing a support system to form a continuous ground support structure (Najafi 2013). The primary difference between utility tunneling and the previous tunneling methods is that the liner remains static after installation instead of being pushed forward by jacks. Many different means and methods are associated with this type of tunneling.

1.4.5.1 Method The utility tunneling method involves two steps:

- Excavation and installation of the primary support, and
- Installation of the product pipe and grouting of the secondary support/liner system.

In the construction phase, two shafts are constructed: the entry shaft and the exit shaft. The tunnel face typically is supported to protect the workers doing the excavation or to secure the tunnel excavation face against collapse or overexcavation, which could result in surface settlements. The face may be supported by boards placed in the tunnel shield, sand shelves, or a mechanized excavator, such as a tunneling machine or TBM. The shield or excavating system is advanced by jacks pushing against a reaction provided by the installed liner at the back of the shield or the machine or by an array of friction brakes pressing outward and onto the formation material. The excavated soil is transported from the excavation face, and the primary support system is installed. These processes (excavation, shield advance, spoils removal, and support system installation) are repeated until the tunnel reaches the exit shaft. The excavation can be achieved by hand digging, mechanical excavation methods, TBMs, or a combination of these methods. The excavated material can be removed by a vacuum extraction system, muck carts or skips, an auger system, or a combination of these systems (Najafi 2013; Read 2004). The liner support system generally consists of steel ring beams and wood lagging or of steel liner plates.

After the complete installation of the primary support system, the carrier pipe (made of any acceptable pipe material) is installed inside the tunnel. The carrier pipe can also be cast-in-place concrete (plain or reinforced).

1.4.5.2 Advantages and Disadvantages The utility tunneling method is used as an alternative to the other personnel entry–sized pipeline installation methods. The method is used for installations ranging from 48 to 160 in. (1,219 to 4,064 mm) in diameter, with sufficient accuracy for gravity applications. The method is generally not restricted in length and in many cases can handle a wide variety of ground conditions, should subsurface variability be an issue. This method requires a high level of skill and is usually a slow process. The method also requires a considerably high capital investment should TBM machines be used.

1.4.6 Pilot Tube Method

NASTT defines the PTM as a multistage method of accurately installing a product pipe by use of a guided pilot tube, followed by upsizing to install the product pipe (NASTT 2013). The *ASCE Manual of Practice for Pilot Tube and Guided Boring Methods* is currently under preparation, with publication expected in 2017. See Chapter 2 for more detail on this method.

REFERENCES

ASCE. (2005). *Pipeline design for installation by horizontal directional drilling.* Reston, VA.

ASCE. (2013). "Overview: Executive summary." <http://www.infrastructurereportcard.org/a/#p/overview/executive-summary> (Jul. 15, 2014).

ASCE. (2015). "Standard design and construction guidelines for microtunneling. "*ASCE/CI 36-15*, Reston, VA.

Atalah, A. (1996). "Design of microtunneling and jacking pipe." *Pipeline Crossings 1996*, L. F. Catalano, ed., ASCE, New York, 395–402.

Atalah, A., and Kariuki, J. (2009). "Cost comparison between horizontal directional drilling and open-cut construction methods in Nairobi, Kenya." *ICPTT 2009*, M. Najafi and B. Ma, eds., ASCE, Shanghai, 1369–1378.

Atalah, A., Iseley, T., and Bennett, D. (1994). "Estimating the required jacking force." *Proc., No-Dig 94 Conf.*, NASTT, Dallas, TX.

Bennett, D., and Ariaratnam, S. (2008). "Horizontal directional drilling good practices guidelines." HDD Consortium and NASTT, Cleveland, OH.

Foster, R. (2009). "Auger boring through hard rock: Overcoming the challenge." <https://www.trenchlessinternational.com/2009/05/07/auger-boring-through-hard-rock-overcoming-the-challenge/> (Jul. 15, 2014).

ISTT. (2013a). "Auger boring." <http://www.istt.com/guidelines/auger-boring> (Nov. 22, 2013).

ISTT. (2013b). "Glossary." <http://www.istt.com/glossary?letter=M> (Nov. 22, 2013).

ISTT. (2013c). "Why trenchless? ('no-dig')." <http://www.istt.com/why-trenchless-no-dig> (Nov. 22, 2013).

Long, D. (2006). "Efficient excavation of small diameter utility installations in hard rock." Tunnelling Association of Canada, Richmond, BC, Canada.

Martin, R., and Grolewski, B. (2011). "Auger boring—A historical review of techniques and applications." *Proc., No-Dig Show 2011*, Paper B-2-01, NASTT, Washington, DC, 1–11,

Mears Group Inc. (2016). "Horizontal directional drilling: Marine crossings." <http://mearshdd.net/hdd/marine-crossings/> (Oct. 9, 2016).

Nada Pacific Corporation. (2013). "Microtunneling." <http://nadapacific.com/microtunneling.html> (Jul. 2, 2014).

Najafi, M. (2013). *Trenchless technology: Planning, equipment, and methods*, McGraw-Hill, New York.

NASTT. (2013). "Glossary of terms and definitions." <http://www.nastt.org/glossary/T?page=2> (Sep. 15, 2013).

Orton, C. (2010). "Pipe ramming for oil and gas pipeline applications." *Proc., No-Dig Show 2010*, NASTT, Chicago.

Osborn Engineered Products. (2016). "Products." <http://www.osborn.co.za/book/export/html/34> (Oct. 9, 2016).

Pipe Jacking Association. (2015). "Pipe jacking." <http://www.pipejacking.org/Resources/PJAintro%202014%20V1%20Final.pptx> (Oct. 9, 2015).

Read, G. F. (2004). *Sewers: Replacement and new construction*, Butterworth-Heinemann, Burlington, MA.

Salem, O., Najafi, M., Salman, B., Calderon, D., Patil, R., and Bhattachar, D. (2008). "Use of trenchless technologies for a comprehensive asset management of culverts and drainage structures." Midwest Regional Univ. Transportation Center, Madison, WI.

Stein, D. (2005). "Practical guideline for the application of microtunnelling methods." Stein & Partner, Bochum, Germany.

TerraMacs. (2015). "Surface launched horizontal directional drilling." <http://www.terramacs.co.nz/trenchless_intro.html> (Oct. 9, 2015).

U.S. Fish and Wildlife Service Midwest Region. (2013). "Appendix J: Horizontal directional drilling." <http://www.fws.gov/midwest/Endangered/permits/hcp/nisource/2013NOA/pdf/NiSourceHCPfinalAppndxJ_HDD.pdf> (Jul. 15, 2014).

Veidmark, A., and Sivesin, C. (2009). "The small boring unit: A ground-breaking alternative to hand-dig installations in hard soil conditions." *Proc., Int. No-Dig Show 2009*, NASTT, International Society for Trenchless Technology, Toronto, ON, 1–7.

CHAPTER 2
HORIZONTAL AUGER BORING SYSTEM

Chapter 2 introduces the HAB system and covers

- Different components of the HAB system,
- Recent innovations in the HAB machine,
- Variations of the HAB system, and
- Considerations and limitations of the HAB system.

2.1 SYSTEM COMPONENTS

The HAB system components discussed in this section include the HAB machine, the steering kit and grade indicator, the cutterheads, the adapter kit, and the augers used for spoils ejection.

2.1.1 HAB Machine

There are four major components of the HAB machine (Fig. 2-1): the power pack, the master casing pusher, the base push unit, and the master track. The power pack contains the engine, the clutch, the transmission, the gearbox, the final drive, the spoils paddles, and the chuck. The power pack also includes the fuel tank, the hydraulic oil tank, the filters, the hydraulic pumps, and the operating controls. The master casing pusher houses the casing adapter and the spoils ejector. The master casing pusher distributes the thrusting force of the hydraulic cylinders to the casing and forms the outside of the spoils ejector system. The casing and auger are secured within the casing pusher housing. The base push unit applies the thrust needed to advance the casing and includes the base frame unit, the header wall, part of the power pack (the spoils ejectors, the front drive, and the

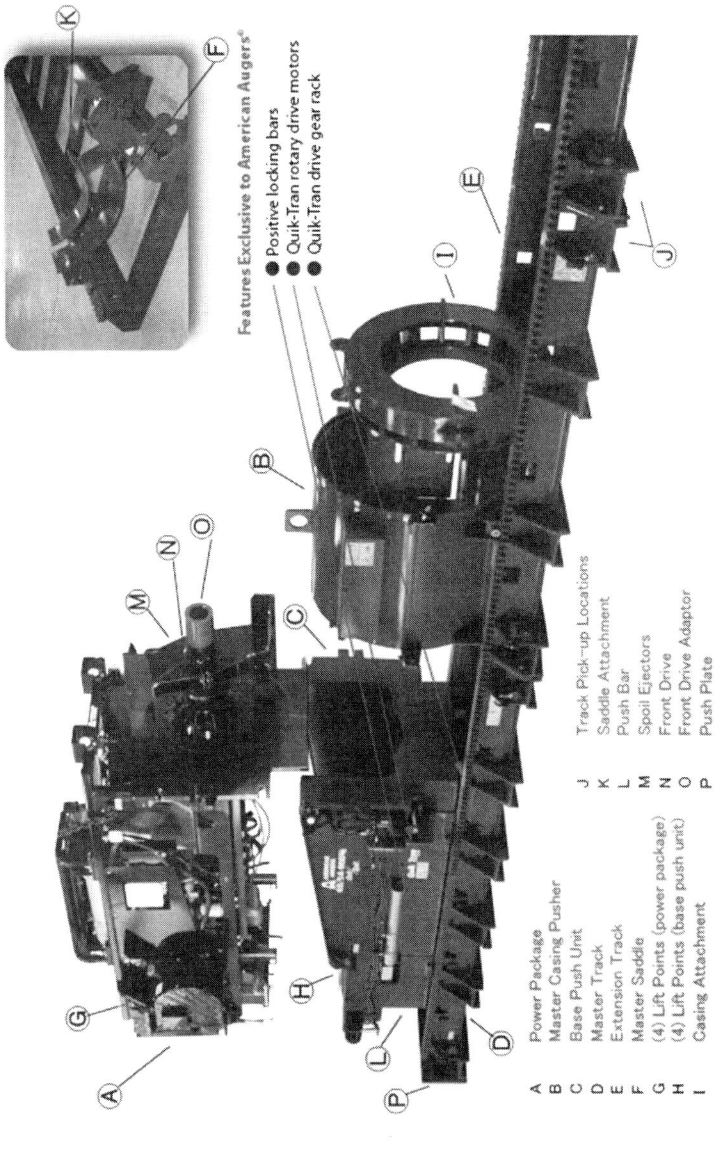

Fig. 2-1. Components of a typical HAB machine
Source: American Augers, unpublished data, 2016

front drive adapter), the thrust cylinders, the push bar, and the operator's station. Thrust generated by the unit can vary from 60 to more than 800 tons (54 to 726 Mg).

The track is composed of three major sections: the master track, the track extension, and the thrust/push/back plate. The HAB machine sits on the master track, which is the rearmost track section. The track extensions are added according to the length of the casing being installed. The casings can be of standard 10-, 20-, or 40-ft lengths (3, 6, or 12 m). In some instances, the tracks can be anchored using rods that are driven into the ground to offer thrust reaction resistance when the boring process begins. The push plate is bolted onto the rear end of the master track and pushes against the thrust block, transferring the force from the hydraulic cylinders to the thrust block, which then transfers the force to the surrounding soil, which provides the reaction force.

2.1.2 Steering Kit and Grade Indicator

Various methods can be used to control line and grade. Each method requires a detector to locate the cutterhead and an articulated section to permit either up–down or both up–down and side-to-side adjustments. The available methods include a mechanical line and grade control head, an electronic line and grade control head, a laser-guided tunnel attachment, and a walk-over system for line guidance combined with water level for grade monitoring. Section 6.10 of this manual has more detail on the water-level indicator and laser-guided systems. The controlled boring system (CBS) and the on-target steering system is described in greater detail in Section 2.2.1.

Steering adjustments in the auger boring process can be achieved using the steering head with articulated joints at the lead, similar to that shown in Fig. 2-2. Since the steering system and the casing pipe are relatively rigid, it

Fig. 2-2. Steering head with articulated joints

is important to periodically monitor the operation and make the necessary adjustments slowly and early to guide the bore.

Another common practice for line and grade control is the use of manual shoes that are installed and adjusted by personnel entry after the augers have been removed. To keep the casing on the intended line and grade, the shoes push the casing in the opposite direction as the bore progresses.

2.1.3 Cutterheads

A wide variety of cutterheads and cutting attachments can be used with the HAB machine to adapt to a wide variety of subsurface conditions. These cutterheads and attachments include all-purpose heads with wing cutters, Christmas tree rock cutters, flat face rock cutters, flat face dirt cutters, combination cutterheads, and other cutting attachments, such as the small boring unit (SBU). All-purpose heads are the most commonly used type. The selection of a cutterhead is dependent on the anticipated subsurface conditions to be excavated. All-purpose cutterheads can handle a relatively wide range of soil conditions but become inefficient as conditions reach the limits of high and low strength and saturation. In these conditions, specialized cutterheads become more efficient, but they also limit the range of materials that can be economically excavated. Once a cutterhead has been selected, the key to the success of any auger bore is the prediction of the penetration rate, which is a function of balancing the excavation rate and the spoil removal rate. It does no good to excavate quickly but not remove the spoils, as the auger system will choke. The selection of the right cutterhead for the subsurface conditions and the prediction of the production rate are critical to the success of the boring operation.

Cutterheads may be fitted with wing cutters (devices attached to the cutterhead that open and close). When the cutterhead is rotated clockwise, the wing cutters open to provide an active overcut of the borehole. Overcut is typically achieved by welding a collar onto the outside of the lead casing to allow the casing to be jacked with less skin friction.

Overcuts are used to reduce friction. Typical situations in which overcuts are used include in dense and hard soil or in very soft and squeezing cohesive soils. Wing cutters, highlighted in Fig. 2-3, are used only in stable soil conditions and are never used with the cutterhead inside the casing. The wing cutters can be adjusted to control the amount of overcut. The standard overcut is 0.75 in. (20 mm) when a steering head is not being used and 1 in. (25 mm) when a steering head is being used. When the cutterhead is rotated counterclockwise, the wing cutters close so that the cutterhead is able to slide back inside the casing for auger removal and servicing. The wing cutters must be set to prevent overexcavation at the bottom of the casing as

Fig. 2-3. Cutterhead with wing cutters (circled)

overexcavation may cause the bore to drift in a downward direction. Overexcavation can be prevented by keeping the cutterhead centered. Centered operation can be accomplished by using new or built-up augers in the lead section of the casing. The use of a worn auger in the lead section will allow the head too much freedom and increase the likelihood of the wing cutter pattern being erratic, which often results in overexcavation below the casing invert and the subsequent diving of the casing. Some of the available cutterhead configurations are presented in Fig. 2-4.

2.1.4 Adapter Kit

The adapter kit includes casing-and-saddle adapters, which are critical when using a machine of a specific size to install smaller-sized casings. For instance, for a machine designed to install casings up to 54 in. (1,370 mm) in. diameter, 24-in. (610 mm) casing-and-saddle adapters could be used to install a 24 in. casing while utilizing that same HAB machine. Fig. 2-5 shows a McL-54/60, which is capable of installing casings up to 60 in. (1,525 mm). To install a 24-in. steel casing, 24-in. casing-and-saddle adapters were used with the machine. The casing-and-saddle adapters add flexibility to the HAB machine by making it possible to use one machine to install casings of various diameters. However, the cutterhead and the augers must be of the same size as the casing being installed.

Fig. 2-4. Different cutterhead configurations

Fig. 2-5. Casing-and-saddle adapters used to install a 24-in. casing while using a large-diameter machine

2.1.5 Different Sizes of Augers

Typically, auger size (diameter) is based on the diameter of the casing being installed, and the auger should be designed to fit into that casing. It

Fig. 2-6. Full-face auger sections in casing

is recommended that all bores be made with a complete string of full-size auger sections (augers designed to fit the inside diameter of the casing), similar to those in Fig. 2-6. However, under conditions in which the auger loading is light, such as with slow excavation rates in harder ground, and in which the spoils move easily in the casing, lead sections of a full-size auger can be followed by smaller-diameter sections. A smaller-diameter auger should never be used in the lead section of the casing without a spoil guidance system to direct spoils into the augers. This process involves a special, nonstandard cutterhead assembly.

The step-down of auger size should not exceed one-third of the casing size in two step-down sizes. When this recommendation is neglected, problems normally occur. Smaller auger sizes can decrease excavation efficiency, as not all the spoils may be removed from the casing. This situation results in a mismatch between the excavation rate and the required spoils removal and requires the auger to rotate more revolutions to remove the volume of soil being excavated at the face.

Bending and torque in smaller auger systems or in the removal of certain cohesive soils that are "sticky" must also be considered when using smaller/undersized augers. The undersized auger is no longer laterally supported by the casing, and it is therefore possible that the auger string will bend. Also, a smaller auger has a smaller shaft and will have more windup from the same torque loading than the full-size auger with a larger shaft. Torque windup pulls the cutterhead back toward the casing and could cause the wing cutters to contact the casing (further increasing torque and resulting in damage) or break the auger string.

2.2 RECENT INNOVATIONS IN HAB

2.2.1 On-Target Steering System

The on-target steering system is a control head designed to simplify the challenges associated with the traditional water-level and laser systems used to monitor and steer the HAB machine.

The control head contains two halogen lights, shown in Fig. 2-7, which are enclosed on its back side. The two halogen lights project twin parallel straight beams of light that enable the head's lateral position, relative to the target, to be monitored from the jacking pit. The steering head in Fig. 2-8

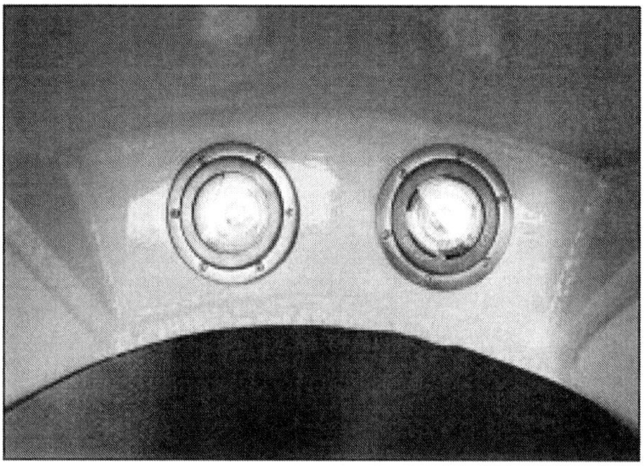

Fig. 2-7. Twin line projection halogen lights
Source: McLaughlin (2013)

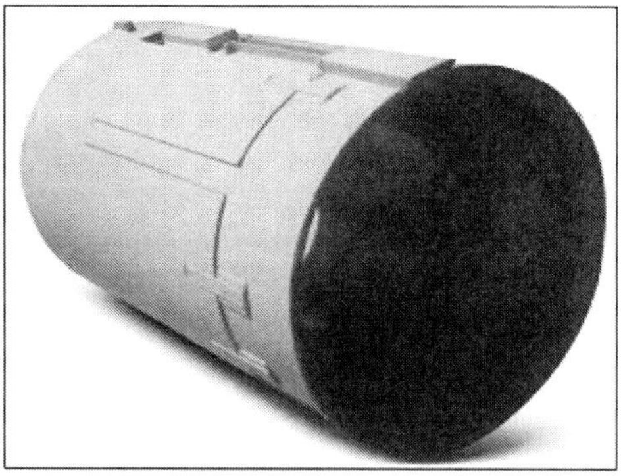

Fig. 2-8. On-target steering head
Source: McLaughlin (2013)

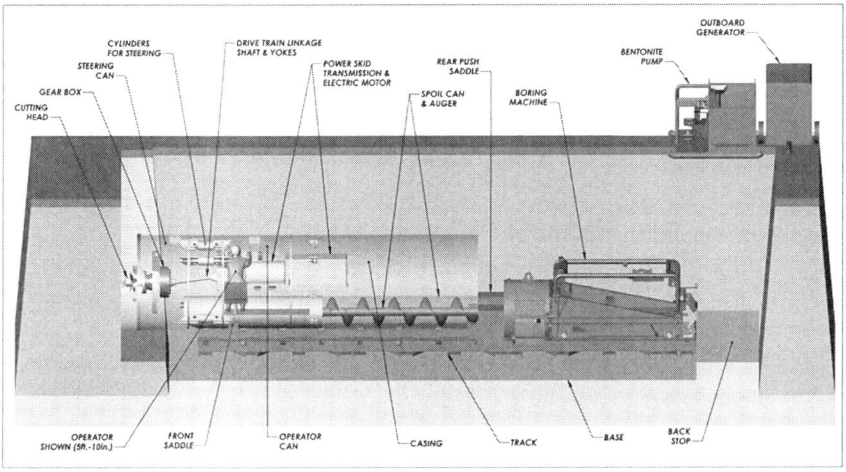

Fig. 2-9. Boring machine tunnel attachment

contains hydraulic actuated flaps that can be opened and closed by the operator to steer the head for both line and grade control (McLaughlin 2013).

2.2.2 Boring Machine Tunnel Attachment

The BMTA, shown in Fig. 2-9, is appropriate for large-diameter installations (>48 in.) and in situations in which line and grade are critical. The BMTA has a self-powered cutterhead that excavates the face. Electric controls provide for controllability of line and grade, and the HAB machine provides the thrust and enables spoils removal through the invert augers (Barbco 2008).

2.2.3 Safety Features

Various safety features have been added to the HAB machine:

- Kill sensor switch: If the cutters engage an obstruction, such as a rock or boulder, while the machine is applying torque, the force can overturn the HAB machine in the shaft and potentially cause tremendous damage, including loss of life. The kill sensor switch was developed to sense an upcoming rapid torque increase upset, trigger the shutdown of the machine, and pull back the clutch, essentially preventing the HAB machine from overturning.
- Remote control of the HAB machine: This improvement enables remote start-up and shutdown of the machine, turns the transmission to automatic, and allows the remote switching of gears. This feature permits an operator to stand in a safer area away from the HAB unit.

- Relocation of the operator platform: To protect the machine operator, some manufacturers have moved the machine operator's platform (operator station) from the traditional location on the side of the HAB machine to the back of the machine. In the event that the HAB machine flips over, the operator will not fall under the machine or be pinned against the pit side.
- Widening of track and machine base: To bring greater stability and avoid rollovers, some HAB machine manufacturers have widened the width of the track and made a wider machine base to lower the center of gravity of the machine (American Augers 2012).
- Replacement of hook rollers with a positive lock system: Originally, the hook rollers that locked the machine into place on the tracks were set manually, which increased the safety risk (if personnel forgot to set the hook rollers). Now, the locking mechanism automatically secures the HAB machine to the tracks, increasing the safety of the operation and the crew in the shaft.
- Use of an operator-tethered kill switch: A lanyard is connected to the kill switch and then goes around the arm of the operator. If the operator gets off the HAB machine by any means, including falling off, the lanyard triggers the switch, which turns off the HAB machine. This measure is used in addition to the emergency shutdown switch at the operator station (American Augers 2012).
- Door flap on the spoil ejection chute: The door acts as an energy absorber and deflector to direct spoils that can rapidly exit the paddles downward instead of up and in the areas often occupied by a laborer removing spoils from the ejection chute.
- Hydrostatic transmissions that permit automatic stoppage of machine advance: When the torque rises faster than a set rate, which could flip the HAB machine should the auger lock onto an obstruction, these hydrostatic transmissions will automatically stop the machine from advancing.

2.3 VARIATIONS OF HAB

The HAB machine has essentially not changed from its original design in the 1930s; however, significant advances in the cutters and steering attachments have enabled HAB machines to economically excavate a wider range of subsurface conditions and sizes while using a single power pack. Cutting attachments include devices such as the SBU, which enables installations in rock. A significant effort has been made to improve HAB machine steering, which had been a significant limitation associated with HAB and often required the installation of larger than necessary casings to permit the realignment of carrier pipes to design grades.

The traditional HAB method allows for grade accuracy but has limited capability to steer left and right. Steering the machine in the traditional system involves pulling the augers out and sending workers to the face to make adjustments. In the last few decades, however, new methods for installing pipes on line and on grade have had an impact on the HAB market. Specifically, the development of the PTM (also known as the GBM) enables precise boring in displaceable soils. The pilot tube can be used in conjunction with HAB to expand the HAB method's alignment capabilities.

Other advances include the development of steering heads that allow the operator to make corrections to the line and grade without having to remove the augers. These systems use externally mounted screws to deflect an articulated steering head and therefore increase the accuracy of the installation without the need to pull and reinsert the auger string.

2.3.1 Small Boring Units

The SBU is a small-diameter cutterhead and thrust-bearing assembly that attaches to the front of the casing. The early units were powered by the HAB auger string system and used the auger string to both drive the cutterhead and remove the spoils. This attachment has extended the capabilities of the HAB machine to include excavation through medium to hard rock conditions and mixed ground conditions. The attachment can efficiently cut rock with a UCS greater than 4,000 psi (28 MPa) (Long 2006) and has been used to cut rock exceeding 25,000 psi (172 MPa). The SBU broadens the capabilities of the HAB machine for easier and faster boring through hard rock in installations ranging from 24 to 78 in. in diameter (Veidmark and Sivesin 2009). The typical setup is similar to the typical auger boring setup, except with a different boring head. The limitation of this technique for harder rock is in the jacking system capacity to apply sufficient force on each cutter and permit an optimal excavation rate.

The SBU is ideal for crossings in medium to hard rock conditions under railroads, highways, runways, and streams, where a steel casing is required to protect the carrier or product pipe. SBUs are best suited for full-face rock but are not suitable for use in soil or under the water table.

Three types of SBUs have been developed over the years: the original small boring unit-auger (SBU-A), the motorized small boring unit (SBU-M), and the small boring unit rockhead (SBU-RH). However, the SBU-RH is not used in concert with a HAB machine and cannot be regarded as a variation of the HAB-powered system; therefore, the SBU-RH is not discussed in this manual.

Fig. 2-10. SBU-A cutterhead attachment (left) and a back view of the unit from inside the casing (right)

2.3.2 SBU-Auger

The SBU-A was the first SBU to be introduced, and it can be used with any conventional HAB machine to install casings ranging from 24 to 72 in. (610 to 1,830 mm) in diameter. The SBU-A utilizes a full-face auger for spoils removal. The longest drive to date for the SBU-A is 600 ft (183 m). This drive will likely remain the upper limit for the machine. Fig. 2-10 shows an SBU-A cutterhead attachment fitted with single disc cutters. The image to the right shows the back side of the SBU-A, which has a centrally placed hex shaft that connects to the full-face auger.

The SBU-A shield is typically welded to the front of the 24- to 72-in. (610- to 1,830-mm) steel casing pipe within which the auger runs. The auger's hexagonal shaft connects to the cutterhead's drive shaft, and the HAB machine rotates the augers along with the cutterhead. As the pipe is jacked forward, the thrust force from the pipe is transferred to the SBU shield, then to the main bearing, and finally into the cutterhead. The jacking force provides thrust to the rock cutterhead to load the cutters, which in turn excavate the rock. As the head is rotated, the rock is fractured and the chips fall to the bottom and are pushed by the rotating scrapers through the cutterhead to the auger. The auger pulls the cuttings through the liner pipe to the jacking pit.

The SBU uses manually adjustable stabilizer pads to reduce/prevent "wobbling" of the cutterhead and steer the casing. During the installation, the augers are removed periodically and personnel entry is required to adjust the stabilizer pads to keep the bore on the correct alignment.

2.3.3 Motorized SBU

The SBU-M can be used for installations ranging from 48 to 78 in. (1,220 to 1,980 mm). The use of a smaller auger system to remove spoils permits

Fig. 2-11. Side view of a typical SBU-M and HAB machine
Source: Fuerst (2012)

the remaining space inside the casing to be used for laser-based steering, hydraulic or electric power to the cutterhead motor system, and entry to service the excavator head. The SBU-M enables personnel entry and longer drives than the SBU-A does and is much more accurate because of the laser-guided technology and the articulating front shield, which enable steering. The cutterhead is driven directly by an electric or hydraulic motor through a gear reduction box using a variable frequency drive (VFD), which makes it possible to achieve greater lengths than are possible with the SBU-A. A small invert auger or vacuum system is used to transport the spoils from the face of the excavation to the jacking shaft (Fuerst 2012). Fig. 2-11 shows a typical SBU-M and HAB machine that is used to provide the thrust and remove the spoils.

2.3.4 Suitability of the System for Various Ground Conditions

Subsurface conditions play a significant role in the performance of the SBU. Ground that is too hard or too soft can hinder the SBU's progress. Very soft rock, with a UCS less than 3,000 psi (21 MPa), can clog the cutterhead and require a slower rotation and advance, particularly if groundwater is present. Very hard rock, with a UCS greater than 25,000 psi (172 MPa), requires greater thrust loading on the disc cutters and can also slow progress and increase cutter changes. Thus, the available jacking force limits the economical excavation of higher strength materials. The system functions best in relatively uniform ground with a relatively narrow range of variation in strength.

SBUs are best suited for full-face rock. They have been used in mixed-face conditions and in full-face soil application, but such use is not recommended. SBUs operate above the groundwater and work well in solid rock or cemented cobbles. SBUs are not suitable for use in soils.

The primary limitation associated with the SBU is the feasible diameter and length. The SBU-A can reasonably function in rock with a UCS up to 25,000 psi (172 MPa). The SBU has, however, been used to bore through granite compressive strength as high as 38,000 psi (262 MPa) through the use of 11.5-in. (290-mm) cutters on a 48-in. (1,220-mm) SBU, but this scenario is

Table 2-1. Suitability of Ground Conditions When Using an SBU

Ground Conditions	Applicability	Comments
Weathered to unweathered rocks	S	Up to 25,000 psi (172 MPa) of rock UCS
Soft to hard cohesive soils	U	Not a suitable technology for full-face soil applications
Medium to dense sands and silts above groundwater	U	Not a suitable technology for full-face soil applications
Very loose to loose sands and silts above groundwater	U	Not a suitable technology for full-face soil applications
Sands and silts below groundwater	U	Not a suitable technology for full-face soil applications
Cobbly and gravelly ground	M	Required to be in a cemented matrix, free of groundwater
Mixed face of soil and rock	M	Transition from soil to rock and then back should be limited to slightly cemented soils with limited cohesion

Note: S = suitable; M = marginal; U = not recommended.
Source: Martin and Grolewski (2011).

not typical. The limits associated with torque have many variables. Table 2-1 outlines the suitability of SBUs to various ground conditions.

2.3.5 Pilot Tube Method

The PTM is a multistage method of accurately installing a product pipe through the use of a guided pilot tube, followed by progressive upsizing to install the product pipe (NASTT 2014). This advancement in auger boring technology dates back to 1995, when the PTM was first introduced in the United States (Boschert 2006). The PTM originated in Europe as a way to use trenchless techniques to install 4- to 6-in. (100- to 150-mm) house connections. Today, the technology is applicable to installations with pipe diameters up to 48 in. (1,220 mm) and drives of approximately 400 ft (120 m) in length. The PTM, also known as the GBM, is a hybrid of HDD,

Table 2-2. Installation Processes When Using the PTM

Three-Phase Process	Two-Phase Process
1. Pushing the pilot tube, 2. Reaming the hole while pushing the auger casing, and 3. Installing the product pipe, which pushes out the casing.	1. Pushing the pilot tube, and 2. Reaming the hole while pushing the product pipe.

microtunneling, and HAB methods. The steering head and pilot bore excavation, followed by multiple passes of hole openers, is similar to the HDD methods, and the guiding system is similar to that used in microtunneling. The spoils removal system is similar to that used in HAB.

2.3.5.1 Components of the PTM The system consists of a traditional jacking frame that advances the pilot tube, casings, and the product pipe. The system utilizes a specially designed theodolite guidance system to guide the installation of the pilot tube. Accurate pipeline installation is achieved through video monitoring surveillance of an illuminated target using the theodolite. Pilot head steering is accomplished by aligning an angled pilot head for the desired course and thrusting forward. Auger flights are used to remove the spoils from the face of the excavation and move them back to the launch shaft for removal.

2.3.5.2 Operation of the System Pipe installation using the PTM can be a three-phase process or a two-phase process (Table 2-2). The pilot tube can be launched either by using a dedicated jacking frame staged in a small 8-ft-diameter (2.4-m) jacking shaft or as an attachment to a conventional HAB machine.

- Three-phase process: The first step is the installation of the pilot tube on the target line and on grade within the specified tolerance. During the installation process, the soil is displaced by the slant-faced steering head. The pilot tube is then directed on line and on grade by rotation during advancement. The hollow stem of the pilot tube provides an optical path for the camera to view the LED target, located in the back of the steering head, and displays the head position and steering orientation. Once the steering head reaches the reception shaft, the theodolite, the video camera, and the monitoring guidance system are no longer needed and are removed from the jacking pit. The pilot hole establishes the centerline of the new installation; the remaining process follows that path.

The second step is to follow the path of the pilot tube with a larger diameter reaming head. The hole enlargement process may take one pass or require progressive enlargements, with the final cut typically slightly larger than the product pipe. The front of the reaming head is screwed to the last pilot tube in the same manner that the pilot tubes fasten to one another. Following the reaming head are auger casings of the same diameter, which transport the spoils to the jacking shaft for removal. Depending on the soil type, a muck bucket or vacuum truck may be used to remove the spoils from the bottom of the jacking shaft. This step is complete when the reamer and auger casings reach the reception shaft and all spoils have been removed.

The third step is to install the product pipe, which replaces the auger casings. The product pipes push the auger casings into the reception shaft, where they are removed one at a time. There are no spoils to be removed in this step since the product pipe has the same outside diameter as the auger casings.

- Two-phase process: An alternative process has been used that combines the second and third steps of the three-phase process. In this approach, a reaming head funnels the excavated material into the auger casings, which are coupled together inside the product pipe. On completion of the product pipe installation, the auger casings are retracted from the inside of the product pipe into the jacking shaft or the reception shaft.

PTM allows contractors to install multiple sizes of sewer lines while using the same set of auger casings. However, as pipe diameters increase, a larger amount of soil must be transported through the casings. PTM equipment manufacturers should be consulted to determine which system is best suited to specific project conditions.

In the United States, the PTM is used to install concrete, clay, polycrete, CCFRPP, and steel pipe ranging in diameter from 4 to 48 in. (100 to 1,220 mm), with a drive length of up to approximately 300 ft (91 m). A few recorded drives have been greater than 500 ft (150 m), but the length is dependent on the subsurface conditions and the pipe diameter. The productivity of the pilot tube installation has been reported to be as high as 200 ft (60 m) per 8-h shift (Martin and Grolewski 2011). The preferable subsurface conditions include displaceable soils, such as dry sands and clays. The method has also been successfully used in displaceable wet sands and soft soils. Rock and soils with an N-value (blow count) greater than 50 cannot be displaced and therefore impose a limit on the use of the PTM. However, some recent advances in technology allow for installations in soils with blow counts above 50 and in solid rock. In addition, grounds with cobbles and boulders cause the uncontrolled lateral displacement of the steering

Table 2-3. Suitability of Ground Conditions When Using the PTM

Ground Conditions	Applicability	Comments
Weathered to unweathered rocks and very dense and hard soils with N-value of 50 blows per foot (305 mm) or greater	U	Not possible at this time; however, air hammers are being tested to make installation feasible
Soft to hard cohesive soils	S	Soils must be displaceable and have an N-value less than 50 blows per foot
Medium to dense sands and silts above groundwater	S	Soil must be displaceable
Very loose to loose sands and silts above groundwater	S	Lubricant is required
Sands and silts below groundwater	S–M	Satisfactory performance with heads less than 10 ft (3 m); marginal performance over 10 ft
Cobbly and gravelly ground	M–U	Probe cannot bypass obstructions that have an N-value greater than 50 blows per foot or ground that excessively deflects the pilot tube

Note: S = suitable; M = marginal; U = not recommended.
Source: Martin and Grolewski (2011).

head and therefore render the method ineffective without frequent realignment, which may not be possible. Accuracy with the method is within ±0.25 in. (6 mm) at 300 ft (90 m). The accuracy depends on the capability of the theodolite and the operator's skill.

2.3.5.3 Suitability of the System for Various Ground Conditions The PTM has been used in weak soils where open-cut and auger boring methods have not been successful. The system works well in displaceable soils with an N-value greater than zero (see Table 2-3 for the applicability of PTM to various ground conditions). Large cobbles and boulders can

Table 2-4. Summary Conditions for Selected Auger Boring Tunneling Methods

Ground Conditions	HAB Machine	PTM	SBU	Comments
Soft to very soft clays, silts, and organic deposits	M–S	M–S	U	Steering systems, if used with a HAB machine, are not effective because of a lack of soil reaction forces
Medium to very stiff clays and silts	S	S	U	Soils must be displaceable for PTM
Hard clays and highly weathered shales	S	S	U	N < 50 blows per foot with PTM
Very loose to loose sands above water table	M	S	U	Ground modifications may be needed to limit settlements
Medium to dense sands below water table	U	M	U	Dewatering is required for HAB; could be required for PTM
Medium to dense sands above water table	S	S	U	Lubrication required for use with PTM
Gravels and cobbles less than 2–4 in. (50–100 mm) in diameter	M	U	U	Dependent on boring diameter; cobbles and gravel tend to bind the HAB machine
Soils with significant cobbles, boulders, and obstructions larger than 4–6 in. (100–150 mm) in diameter	M	U	M	Dependent on boring diameter; should be limited to 33% of the excavated diameter
Weathered to unweathered rocks and firmly cemented soils	S	U	S	Approximately 25,000 psi (172 MPa) for a HAB machine and SBU

Note: S = suitable; M = marginal; U = not recommended.
Source: Martin and Grolewski (2011).

Table 2-5. Capabilities and Limitations of the Auger Boring Methods

Description	HAB	PTM	SBU	Comments
Min./max. excavated diameter [in. (mm)]	4/96 (100/2,440)	4/48 (100/1,220)	24/78 (610/1,980)	HAB can achieve an excavated diameter of 156 in. with an LDBA/BMTA
Min./max. drive length [ft (m)]	50/600 (15/183)	50/300 (15/91)	50/500 (15/152)	
Installation accuracy (±)	1% of drive length	0.25 in. (6 mm)	1% of drive length	Accuracy can be improved to ±4 in. irrespective of drive length when steering attachment is used with HAB machine or SBU

Note: min. = minimum; max. = maximum. LDBA = large-diameter boring attachment; BMTA = boring machine tunnel attachment.

prevent the pilot tube from advancing. Recent developments, such as lubricants for loose sands, water control reaming heads for wet sands, and testing of air hammers for use in rock, increase the range of subsurface conditions in which the PTM is used.

2.3.5.4 Developments in the System Some of the recent developments that have increased the capability of the technology for dealing with a wider range of soil conditions include hydraulically powered cutter and reaming heads, lubricants for loose sands, and water control reaming heads for wet sands. Advances have also been made to use small-diameter cutterheads to accommodate installation in rock.

The PTM can be used in conjunction with other trenchless methods to provide the degree of accuracy needed for some projects.

2.4 CONSIDERATIONS AND LIMITATIONS OF HAB

HAB can be used on a wide variety of subsurface conditions. When selecting the most appropriate method, consideration should be given to the subsurface conditions and the length of the drive. Table 2-4 is a guideline for selecting the most appropriate method for specific ground conditions on the basis of the historical success of the HAB method in various types of ground. Table 2-5 presents the capabilities and limitations of auger boring methods. More details on subsurface conditions are covered in Chapter 4 of this manual.

REFERENCES

American Augers. (2012). "Auger boring machines." <http://www.americanaugers.com/equipment/auger-boring-machines.html> (December. 17, 2013).

Barbco, Inc. (2008). "Boring machine tunnel attachment." <http://www.barbco.com/download/advanced/BMTA-Pit-Illust_HR.pdf> (Oct. 9, 2016).

Boschert, J. (2006). "Pilot tube microtunneling catching on in the United States." <http://trenchlessonline.com/pilot-tube-microtunneling-catching-on-in-the-united-states/> (Jul. 2, 2014).

Fuerst, T. (2012). "Method statement for 1200 mm (48″) diameter motorized small boring unit (SBU-M)." Robbins Company, Solon, OH.

Long, D. (2006). "Efficient excavation of small diameter utility installations in hard rock." Tunnelling Association of Canada, Richmond, BC, Canada.

Martin, R., and Grolewski, B. (2011). "Auger boring—A historical review of techniques and applications." *Proc., No-Dig Show 2011*, NASTT, Washington, DC.

McLaughlin. (2013). "On-target auger boring steering system." <http://www.mclaughlinunderground.com/products/target-auger-boring-steering-system/> (Nov. 22, 2013).

NASTT. (2014). "Glossary of microtunneling terms and definitions." <https://www.nastt.org/microtunneling_glossary> (Jul. 2, 2014).

Veidmark, A., and Sivesin, C. (2009). "The small boring unit: A ground-breaking alternative to hand-dig installations in hard soil conditions." *Proc., Int. No-Dig Show 2009*, NASTT; International Society for Trenchless Technology, Toronto, ON, Canada.

CHAPTER 3
PLANNING PHASE

One of the major struggles in developing an MOP for HAB is the high versatility of applications to which the HAB technology may be applied. These applications have different challenges and risks and result in the development of a wide range of operator skill sets to work with the given subsurface conditions.

Project complexity and the risk of encountering hazards determine the appropriate level of planning for a successful project. Some crossings are short and located underneath noncritical roads. Often, these installations can be completed with minimal effort in planning, geotechnical investigation, design, and construction and can be installed without grade monitoring and steering. However, some bores are long and located in challenging environments and/or ground, and significant effort in planning, geotechnical investigation, design, and construction may be required to manage the risks. Creating one manual that addresses every possible challenge and risk is impractical because of the wide variety of HAB applications. The potential solutions to various subsurface conditions are also wide ranging; therefore, the judgment of the design and construction teams is essential to managing the risks and challenges for specific project conditions.

Chapters 4, 5, and 6 are written for average projects with average levels of risk and complexity. Therefore, less risky projects can utilize portions of the activities discussed in these chapters; more challenging projects may use all the suggested actions in these chapters, plus any other risk mitigation action mandated by the local site conditions.

The keys to successful planning are understanding the excavation process on the given ground, identifying the hazards that may occur along the path, and managing the risks of occurrence through

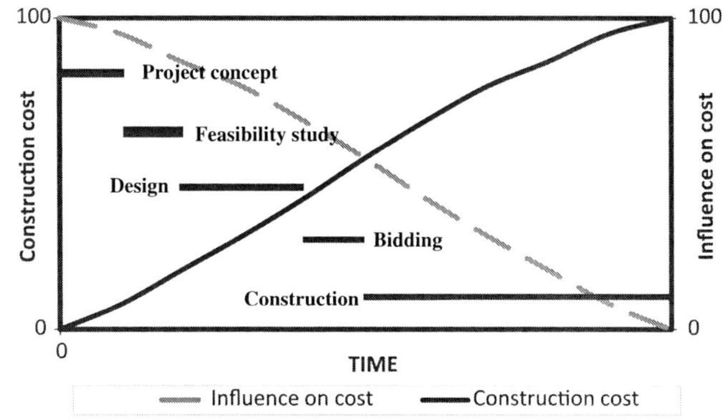

Fig. 3-1. The ability to influence construction cost over time

contingency planning. As is typical for construction projects, the ability to influence the project cost is highest at the planning phase (Project Management Institute 2013), as shown in Fig. 3-1. The planning phase outlines the basic elements required for successful project completion and examines and determines the suitability of HAB for the project. This phase provides sufficient information to enable the design engineer to design the project and develop the drawings and contract documents that the contractors will use to estimate the cost of the project competitively and with appropriate risk allocation.

The HAB planning phase should include some or all of the following elements, as necessary to define and control the project costs, schedule, risks, and means and methods:

- Feasibility study
- Predesign survey and alignment layout
  - Location of the jacking and receiving pits
  - Line and grade tolerances
  - Shaft depth and bore length feasibility
  - Pipe size consideration
  - Subsurface and overhead utilities
  - Preliminary geotechnical investigation to determine suitability of subsurface conditions
  - Site reconnaissance to determine availability and suitability of construction area
- Environmental and social considerations
  - Community impact (noise, dust, traffic, parking, etc.)
  - Environmental impact, such as collateral impact from dewatering, hazardous materials in the ground, excavation failure in sensitive area, and so on
- Regulatory considerations and required permits

The owner selects the procurement method in the planning phase. This method could be traditional design–bid–build, design–build, construction management procurement, or an integrated approach. The procurement method that is selected affects the level of interaction and input from all project participants. In the design–build procurement method, for instance, the construction contractor is involved very early in the process, which could be critical in high-risk and complex projects. The owner may elect to have an advisor to assist in making a decision on the procurement method.

Another important procurement decision is the scope of work of the HAB crossing: Will the owner hire the HAB contractor directly? How and when will the HAB contractor be paid? Will the withholding only be released after the general contractor reaches substantial completion? What contingency funding is available for risk management? The typical scenario is an open-cut project that includes HAB crossings. The owner hires the general contractor, which hires the HAB subcontractor.

3.1 FEASIBILITY STUDY

The feasibility study leads the owner to make the decision to invest in the project and to initiate funding procurement. The basic objective of the feasibility study is to define the project's needs, benefits, costs, and cost/schedule contingencies. Estimates of the cost and benefits in this phase are usually preliminary in nature but are used to determine the need for further studies. The tasks to be completed during the feasibility analysis include the following:

- Identify and define the project needs. This task would include determining whether concrete encasement and a carrier pipe are needed.
- Identify the allowable project budget and duration.
- Explore the possible alternatives to meet the project needs, for example, open-cut, HDD, or HAB methods.
- Define the selection criteria for the potential alternatives: cost, schedule, inconvenience, and so on.
- Select the alternative that best meets the project needs on the basis of the selected criteria.
- Verify the project feasibility in terms of physical practicality, cost, and social and environmental constraints.
- Determine possible hazards and assess if and how the anticipated risks of encountering the hazard are manageable or require additional study during design.

3.2 PREDESIGN SURVEY

The first order of business for planning any project is to define the project requirements for success. Project requirements include the need for casing, carrier diameter, grade specification and tolerance, depth range and tolerance, and project start and end locations. For the successful execution of HAB projects, it is essential to conduct site reconnaissance to scout the site conditions and evaluate potential hazards and challenges that need to be resolved during the planning, design, and construction phases of the project. The investigation of general site conditions may include existing underground and overhead utilities; the potential for obstructions; ground conditions, including groundwater; environmental conditions; conceptual HAB drive lengths; pipe diameters; site access; depth; grade; tolerances; the impact of the project on surface activities; the location of existing/abandoned/proposed utilities; local work and traffic regulations; and right-of-way requirements.

3.2.1 General Survey of Site Surface Conditions

The general site survey is based on local experience, desktop information, and field observations by experienced professionals, including trenchless technology professionals, owners' representatives, and design engineers. It is beneficial to complete a desktop study to collect information from experienced stakeholders and engineers prior to field visits. It is possible to obtain aerial photographs; rough elevations; and information on the site usage, layout of existing features, and general subsurface conditions. Local utility field personnel often have knowledge of local utilities and local subsurface conditions. Local utility or trenchless technology contractors may be willing to share their local knowledge on past projects. The local tax office can provide rough lot ownership information. Local officials and possibly the fire department may be able to provide data on local environmental actions—such as buried tanks or releases of hazardous materials—that could impact the project. Local regulators can provide indications of the permits that may be required and a timeline and data package for obtaining those permits. Local officials can provide input into the importance and political impact of various types of work in the vicinity. This information should be captured and recorded as project background data. The assessment of this information leads to an initial list of potential project issues that can focus the initial site visit observations and the project data and study requirements. Initial site visit observations and initial data gathering may include the following:

- General site character: urban, residential, commercial, rural, and so on.
- Existing overhead utilities relative to possible alignments.
- Indications of underground utilities on the basis of manholes and pavement patches or uneven ground.
- Surface soil and debris that may be indications of potential obstructions and subsurface conditions.

- Surface water observations, including stormwater routing in the area of the alignment.
- Environmental conditions, such as the potential for contamination from industrial or commercial site usage.
- Site access and room for construction staging and shafts, with the resulting possible HAB drive length and ownership of possible worksites and land acquisition requirements for right-of-way and construction easements.
- Site safety concerns for given construction requirements.
- Impact of the project on local surface activities and businesses.
- Location of existing/abandoned/proposed utilities.
- Local work and traffic patterns and critical crossings, such as railroads or environmental features, that can cause special design actions and impact cost and schedule.

This information plays a vital role in determining the feasibility, alignment, gradient, and physical constraints of the proposed HAB project.

Project locations in some flood-prone areas necessitate an assessment of flooding risk, which should include a review of information from FEMA, the National Oceanic and Atmospheric Administration, and the U.S. Army Corps of Engineers (USACE), as well as local knowledge, to determine whether storm impacts are probable, their expected seasons, and their frequency. In some locations, flooding may last for weeks and affect the duration of the project. The ground may become saturated, which can affect the stability of the pits and the boring operation. Because of the high risks involved, it is advisable to schedule auger bore construction jobs during nonflood periods or have contingency plans in place.

3.2.2 Alignment Considerations

The majority of the HAB installations involve a single run that crosses under a road, railway, or airport runway. However, in a case of projects with multiple bore reaches and direction changes, an evaluation of pipeline route alternatives should be considered to determine the optimal alignment, including the trenchless crossings. The optimal alignment is the alignment with the best fit and the fewest challenges. In critical crossings, the selected alignment must be plotted on a plan to evaluate the available right-of-way, the depth of and proximity to existing utilities, the possible shaft locations, the slope, easements, access, the equipment and material delivery routes, and the construction staging areas.

Given that alignments for HAB are straight projections from the jacking shaft to the receiving shaft on either side of the road, railroad, or airport runway, the bore path can be staked or marked along the proposed alignment. The depth of the alignment can vary depending on the geology, topography, existing utilities, and intended purpose for the product pipe.

The use of steel casing for HAB allows high thrust loads for crossings in varying geological conditions, including cobbles and rocky grounds. Although simple but proven methods for steering the casing straight are commonly used, innovative systems and products are commercially available for a higher degree of accuracy in line and grade control. However, it is a basic concept in trenchless construction that, regardless of steering precision, the existing ground reaction to the excavation process controls the accuracy.

It is important that the subsurface investigation program provides substantive information for the engineer to evaluate the possibility of encountering utilities and obstructions, such as boulders, cobbles, rock, timber piles, concrete, bricks, and metal objects, before the pipeline alignment is established.

3.2.3 Traffic Information

All traffic information in connection with the project alignment needs to be collected. Traffic information is obtained to determine the social cost and project impacts should traffic need to be rerouted or physical traffic protection barriers need to be installed for safety. Traffic information should include a site access assessment with local highway or street officials to obtain a project access corridor to a highway that is capable of supporting the project traffic. Any bridges on the access corridor should be capable of supporting the heavy construction equipment. The information should be used to minimize disruptions to vehicle and pedestrian traffic patterns, driveway and business access, and off-street parking.

Since traffic control and construction costs can be calculated for each alternative, these costs can be compared with the relative benefits for each method. The benefits of each method's traffic control strategy may include such things as reduced delay and travel distance. Costs of traffic control and construction need to be reviewed as part of the evaluation and should be readily available as part of the design effort. These costs include an appropriate allowance for traffic control enforcement, when appropriate.

The cost to vehicles of traffic delays should be accounted for in the overall project assessment as this cost can have a significant political impact. The traffic delay cost is the cost of time, in dollars per vehicle, for the delay of traffic through and around construction activities.

3.2.4 Preliminary Geotechnical Investigation

The results of a desktop study that identify potential subsurface issues form the basis and need for additional subsurface studies, including geotechnical investigation. As with all underground projects, experience and good judgment are critical ingredients throughout the site investigation's development, implementation, and interpretation process. An assessment should be made of the desktop-identified subsurface information to determine the risk of facing the hazard and the impact on safety, cost, and schedule should a hazard be encountered. A formal risk evaluation need not be done on every project, but the simple process of developing a cost-effective subsurface investigation program is warranted. The results of this process may indicate that sufficient information is available from the desktop study for the project to go to construction or they may suggest a need for additional investment to verify and quantify certain hazards to eliminate or mitigate the risk.

A qualified engineer should determine the extent of the preliminary geotechnical investigation. Much of the investigation in the planning phase involves a site visit; the review of historic documents; discussions with people who have knowledge of historic land use; and the review of geologic drawings and maps, previously completed geotechnical investigations, soil reports, and any other reports specific to the area of interest. From the desktop study, it will be possible to determine whether there is sufficient information available to identify all the appropriate issues and provide sufficiently accurate data to permit a proper assessment for the design of solutions by both the engineer and the contractor. If the conclusion indicates that the information is insufficient for the risk level and the associated cost of managing those risks, then additional studies and explorations should be completed or adequate schedule and cost contingencies need to be set up by the project owner to fund the premium cost of the contractor encountering and handling the hazards during construction.

Postponing the site investigation to the construction phase of a project significantly increases the risk that the project cost and schedule will be affected, reduces the time available for developing options, and may restrict the mitigation options available, resulting in delays and change condition claims.

3.3 COST CONSIDERATIONS

Various factors influence the cost of a HAB project, including

- Geotechnical conditions;
- Location of the project;
- Existing utilities along the alignment;
- Previous land use;
- Groundwater;

- Complexity of the project;
- Depth of the pits;
- Available working space (especially for the jacking pit);
- Diameter of the pipe;
- Length of the bore;
- Number of bores on the job;
- Mobilization, setup, and moving costs;
- Required accuracy;
- Environmental impact;
- Market conditions;
- Public and community perception;
- Previous HAB success/failure; and
- Traffic volume.

The overall project cost estimate should include the design, construction, operation and maintenance, and disposal costs. The benefits may include, in addition to the planned function, the social and environmental impacts. The project cost should begin with the desktop study and be continually updated as new information becomes available. This information needs to be shared with the project owner and the finance partners to permit proper owner decision making during the project. Initial costing may not have a high degree of accuracy, but at a minimum, placeholders for cost items should be shown and accounted for so that financial surprises are managed during the project.

3.4 ENVIRONMENTAL CONSIDERATIONS AND SOCIAL BENEFITS

Construction work in environmentally sensitive areas, such as wetlands, rivers, streams, natural habitats, public parks, and historic places, involves a lot of effort (because of the legal and regulatory requirements to protect these sensitive areas). Public criticism of surface disruption and administrative burdens tends to increase for projects planned and undertaken in these areas. With the rising concern over pollution caused by construction activities, especially in environmentally sensitive areas, the U.S. Environmental Protection Agency (EPA) and other authorities are creating strict rules about work spaces in these areas. The HAB method can significantly reduce the physical impact of construction in environmentally sensitive areas. Additionally, studies have shown that environmentally sensitive carbon emissions may be significantly reduced by the use of HAB or other trenchless methods.

It is important to avoid environmentally sensitive areas as much as possible, especially when selecting the locations for the entry and exit pits.

Economic costs and benefits, and the probability and impact of a negative event, should be the guiding principles when making this decision. Below are some of the environmental and social benefits of the HAB method over the conventional open-cut method.

3.4.1 Environmental Benefits

The conventional trench-and-cover construction methods for utility installation are often associated with noise and vibrations generated by the operation of construction equipment, excavation, shoring, and the installation and removal of temporary ground support systems. To install buried utility lines, contractors have to cut open existing surfaces, such as pavements, using heavy machinery like jackhammers, backhoes, excavators, and trenching machines. This equipment produces vibrations and noise that often lead to citizen complaints. Additionally, open-cut utility construction tends to produce dust, and the health concerns associated with these construction methods are obvious public nuisances. This problem gets more complicated in sensitive urban areas with schools, hospitals, homes, and businesses. In comparison to the trench-and-cover construction method, the trenchless installation of pipelines by HAB mitigates the negative effects of noise, vibration, and pollution.

3.4.2 Social Benefits

Several social benefits are associated with trenchless pipeline installation. The following are some of the social advantages to the appropriate use of a trenchless installation method, such as HAB, over the open-cut method:

- Smaller construction footprint and reduced pavement removal;
- Reduction in utility conflicts, damage, and service interruption;
- Minimal disruption to pedestrian and vehicle traffic;
- Less disruption to homes, businesses, and commerce;
- Enhanced safety;
- Less excavated material handling and disposal; and
- Smaller carbon footprint.

3.5 REGULATORY REQUIREMENTS

When boring in environmentally sensitive areas, the team must comply with any special environmental regulatory requirements. It is the job of the permitting engineer and the design engineer to educate the regulators and negotiate realistic constructable regulatory conditions for the project. It is then the responsibility of the team to properly price and comply with the given conditions. It is important to comply with all the regulatory requirements for the particular job. Sometimes, excavated soils are contaminated with material that could be hazardous. The health and safety of the working crew and the public must take precedence. The management, use, and/or disposal of the contaminated material must comply with the legal requirements for handling such material; these requirements can result from federal, state, or local area regulations.

In a project in which dewatering is required, the EPA, state, and local agencies may have permitting requirements for discharging the water. Water could be discharged to waterways, storm channels, infiltration or evaporation ponds, and local sanitary sewer or storm drain systems. If the water is contaminated, some level of treatment is required prior to discharge, depending on the constituents and concentrations present (URS Greiner Woodward Clyde 2000).

When crossing under a railroad or highway, the work must be done in accordance with the requirements of the agencies that own and operate those assets (e.g., the railroad company, the state DOT, or the city). It is critical to ensure that the project is compliant with the applicable regulations, which could include

- EPA requirements,
- Occupational Safety and Health Administration (OSHA) requirements,
- National Historic Preservation Act requirements,
- National Pollutant Discharge Elimination System (NPDES) requirements, and
- Federal Endangered Species Act requirements.

In addition to compliance with regulatory requirements, it is important to take every precautionary action to ensure safety and minimize risks to safety, the environment, cost, and schedule.

3.6 SHOW STOPPERS AND PREMIUM COST CONDITIONS

During the planning phase, the owner should identify potential show stoppers and take measures to avoid them or reduce their probability of occurrence and/or have a plan to mitigate their negative impact should they occur.

3.6.1 Potential Show Stoppers

Potentially significant problems encountered in the field include

- Boulders more than one-third the size of the casing and/or nested boulders larger than one-third the size of the casing;

- Mixed-face conditions of soil and solid rock;
- Wet flowing sand;
- Installations below the water table with the potential to flood the machine; and
- Installations that encounter unexpected rapid water influx from any source, such as a saturated layer of sand with perched water.

3.6.2 Premium Cost Conditions

Conditions that increase the project cost include

- Projects that require a high level of accuracy, which may need guidance systems that have an increased cost;
- Hard rock conditions, which require the use of rock boring systems, such as SBUs; and
- Mixed-face conditions, which may require different tooling to complete the drive. If the soil is homogeneous, then the tooling is specific and may not need to be changed during the run. The use of a single tool would not be practical for mixed ground conditions.

REFERENCES

Project Management Institute. (2013). *A guide to the project management body of knowledge*, 5th Ed., Newtown Square, PA.

URS Greiner Woodward Clyde. (2000). "Microtunneling design guidelines." Offshore and Onshore Design Applications Supervisory Committee of Pipeline Research Council International, Inc., Houston.

CHAPTER 4
GEOTECHNICAL INVESTIGATIONS

HAB involves the controlled excavation of various ground conditions, generally above the GWT or in areas where the water table can be actively controlled. The process starts with characterizing the type of geologic material and its possible impacts on the components of a HAB system. The success of the design and construction work for a HAB project depends greatly on matching the earth materials to the system components, such as the cutterhead, the lubrication, and the jacking capacity.

Soils are generally classified into two major groups: cohesive and cohesionless. Cohesive soils contain materials that can form bonds between particles to increase strength and include the fine-grained subgroups consisting of clays, organics, and some silts. Cohesionless soils contain particles that do not bond but rely on intergrain packing and particle contacts to obtain strength. AASHTO and the Unified Soil Classification System are the commonly used soil classification systems based on grain size and behavior. The Modified Burmeister classification system is also used for soil classification. More information about these classification systems is available in most soil mechanics textbooks.

Rock properties include those at the intact rock scale (small scale) and those at the rock mass scale (large scale). ASTM International and the International Society of Rock Mechanics (ISRM) provide specifications and guidelines for the collection of field data to determine the properties and behavior of the rock through rock mass classification systems. Data at both the intact rock scale and at the scale of overall rock mass behavior aid the HAB contractor in selecting auger boring components for projects in rock. In addition, such data help reduce the project risk by identifying zones of potential water influx, ground variability, ground pretreatment needs, and settlement/heave potential.

A common and accepted practice is to conduct geotechnical site investigations in phases, in which each phase provides the information needed for the execution of the next phase of the project. A typical phasing would start with a desktop study, which would be followed by exploratory borings, geophysical investigation, further borings, laboratory tests, and information synthesis in a geotechnical report.

The required extent of the geotechnical investigation is a judgment call from the design and geotechnical engineers and is based on the complexity of the project and the potential hazards. The risk tolerances of the owners of the pipes and of the roadway, railway, runway, or other facility also play an important role in determining the extent of the investigation. For low-risk projects with high risk–tolerant owners, a desktop study may suffice; however, a thorough geotechnical investigation may be conducted on riskier projects with less risk-tolerant owners.

This chapter covers

1. Importance of geotechnical investigations,
2. Risk management from the geotechnical standpoint,
3. Geotechnical investigation programs,
4. Geotechnical properties, and
5. Geotechnical reporting requirements.

4.1 IMPORTANCE OF GEOTECHNICAL INVESTIGATION

Soil and rock are naturally occurring earth materials that have been subjected to various geologic processes. As such, soil and rock present an inherent three-dimensional variability in material properties. The geologic processes tend to segregate and separate these materials into soil deposits or rock formations that have some common characteristics. If this fact is appreciated in advance, the geotechnical engineer can optimize the field investigation program for site-specific geologic conditions and thereby maximize the benefit of the data with respect to budget constraints.

Soils and rocks behave differently during boring operations, and the same soil behaves differently under a range of field conditions (e.g., dry sand above the GWT behaves differently than the same sand below the GWT). Also, the same soil may react differently to the boring process during different seasons of the year. If the expected soil materials are identified and their distribution is determined with a high level of reliability, the geotechnical engineer and the HAB contractor can estimate how the material will behave during the advancement of the HAB machine along a drive. A clear understanding of material properties and the excavation process significantly reduces risks and increases the potential for success. By understanding the reaction of the subsurface material to the excavation process, the HAB contractor can apply suitable tooling and techniques to control the material reaction and maintain stability. Therefore, the anticipated in situ material behavior during HAB guides the selection of the appropriate alignment and depth, construction method, boring system components, support of excavation at the jacking and receiving pits, jacking loads, and lubrication, to name a few factors.

The value of data from a well thought out and properly executed geotechnical investigation program cannot be overstated. Geotechnical considerations can determine the feasibility of most underground and trenchless projects. The geotechnical investigation and analysis guides all parties to identify the inherent hazards and associated risks of underground construction and therefore leads to successful project completion (within schedule and budget) (Peck 1967). The actual subsurface data exploration extends through the construction process as well.

The geotechnical investigation and analysis process guides the owner, engineer, and contractor in the following:

- Successful planning, design, and construction;
- Baseline conditions assessment; and
- Dispute resolution and settlement.

The geotechnical investigation, therefore, guides the engineer to do the following:

- Design the project;
- Finalize the alignment, diameter, slope, and depth;
- Select acceptable construction methods that are constructable for the given condition;
- Identify locations for the jacking and receiving pits and the ancillary structures;
- Confirm the feasibility and constructability of the project;
- Develop specifications, drawings, and bid documents;
- Develop cost estimates and schedules; and
- Anticipate the associated construction hazards and quantify the risks.

The geotechnical investigation guides the contractor to do the following:

- Select the construction equipment and method;
- Estimate the jacking force;
- Determine the hydraulic jacking capacity of the HAB machine;
- Select the cutterhead type, material, and configurations;
- Select the lubrication fluids;
- Design the construction process parameters;
- Design the thrust block and shoring system for the jacking shafts;
- Design the shoring system for the receiving shafts;
- Develop a rational cost estimate;

- Develop a schedule and control the project; and
- Manage the associated construction risks.

4.2 RISK MANAGEMENT AND GEOTECHNICAL INVESTIGATION

A hazard is an event or a thing that can cause harm to persons or their property. "Risk" refers to the probability of encountering an event (hazard) that will impact project safety, the environment, cost, or schedule. The main tool for managing risks is to define and identify the hazards and estimate their probability of occurrence and the resulting impact. The probability is measured on a scale of 1 to 100%, where a probability of 1% means the hazard happens very rarely and a probability of 100% means the hazard happens very frequently. The impact is also measured on a scale of 1 to 100%, where 1% means the risk has very minimal impact and 100% means a catastrophic impact (Alexander and Marshall 2006). Examples of high impacts are loss of life; damage to nearby buildings and infrastructures; and the closing of major highways, railways, waterways, or runways. For projects that have high-frequency and high-impact risks, deep and thorough geotechnical investigation is needed. For example, if the soil is stratified, has high variability in estimated shear strength, and/or includes boulders/cobbles underneath a critical highway, railroad, waterway, or runway, the majority of the soil investigation techniques described in this section should be adopted. However, for projects with low-frequency and low-impact risks, shallower, less invasive, and less costly explorations may be acceptable.

If the probability of a risk is more than 50% and the impact is more than 50%, taking measures to reduce the probability and mitigate the impact of risks during the planning and design phases is usually more economical than dealing with those risks during the construction phase. For instance, although it is possible to remove the augers from the casing and remove an obstacle in large-diameter bores, such intervention takes additional time and effort, which increase cost. It is also risky for personnel to work inside the casing to remove the obstruction. OSHA addresses confined spaces without specifying a minimum size for personnel entry, but diameters smaller than 42 in. (1,065 mm) are challenging and considered small for personnel entry. The choice to abandon a bore drive or excavate to remove obstacles from the surface to complete the crossing adds significantly to cost, delay, and inconvenience to the users of the crossing.

The magnitude and depth of a geotechnical investigation depend on the risks associated with the project. Every construction project has some risks, such as a significant increase in cost and/or duration, failure to complete the project, safety risks, and environmental risks.

The primary cause of high risks in HAB projects is inadequate geotechnical information for the identification and quantification of hazards and the appropriate matching of the HAB machine tooling with the actual ground conditions. In the case of a lack of defined ground condition and a project specification that requires a specific construction means and method, an owner will typically be held accountable for presenting a constructable project, and a contractor may only be required to follow industry standard means and methods for the specified technique (the Spearin doctrine). The more information the engineer and contractor have about the expected subsurface conditions, the better the engineer and contractor can design and the better a contractor can select the appropriate tooling for the HAB machine. Some of the major issues include the presence of boulders/cobbles, mixed-face ground conditions, groundwater conditions and location, noxious gases, softer than expected soils, harder than expected soil/rock, artificial obstructions, and a high level of variability in the soil/rock properties.

Engineering and geotechnical judgment is needed to determine the depth and magnitude of geotechnical investigations required to identify risks with a low probability of occurrence but high impact. The owner and the engineer may decide that, if the hazard materializes and the HAB machine encounters challenging conditions for which it was not designed, the owner will pay for the extra work through change orders as described in the contract documents. For example, if in a specific project area it is rare to encounter cobbles/boulders, the owner may reduce the up-front cost of the geotechnical investigation and pay for the additional cost via change order if cobbles/boulders are encountered. If the project is located in an area with minimal available geotechnical information, increased exploration is recommended. Whether the probability and impact of the risk are high or low, the owner needs to determine and implement every reasonable measure to mitigate the risk, and the engineer should inform the owner of the hazard and the associated risk. Such risk-mitigating measures could include

- Thorough geotechnical investigation and analysis to select a cutterhead that is capable of excavating the soil economically and efficiently along the target alignment and grade without excessive settlement or heave;
- Use of grouting to enhance the soft soil properties;
- Design of the pipeline alignment in a different location;
- Definition of dispute resolution measures in the contract so that claims can be resolved fairly and quickly;
- Preparation of what-if scenarios and the development of contingency plans, including alternative traffic detours; and
- Additional targeted geotechnical explorations.

4.3 GEOTECHNICAL INVESTIGATION PROGRAM

Geotechnical investigation provides data for the engineer to design the proposed project and reliable information to the contractor to determine reasonable means and methods to bid and install the pipe successfully. Program objectives include the following:

- To provide an adequate database for engineering decisions based on in situ and laboratory testing (including geophysical testing, if required),
- To generate a description of in situ soil and rock conditions and their distribution,
- To define the design range of groundwater elevations,
- To develop field and empirical estimates of in situ soil and rock properties and behavior, and
- To determine the potential range of project hazards and risks both to the immediate construction and adjacent properties.

4.3.1 Phases of Geotechnical Investigations

The geotechnical site investigation is usually conducted in phases, and each phase informs and guides the next phase of study. The desktop study provides the focus for the initial exploration program and study. The study consists of acquiring ground truth of the identified hazards and issues through the use of exploratory borings, geophysical investigation, further borings, and laboratory tests. The results are then assessed, and the data and assessment are documented in a geotechnical report.

The first phase is conducting a desktop study to review existing geotechnical information. A simple prefield investigation desktop study that examines publicly available information (e.g., texts and Internet studies) can be extremely useful. The following activities can be part of the desktop study:

- Field reconnaissance visits;
- Remotely sensed satellite imagery, including aerial photography;
- Local topographic and geologic maps;
- Geologic reports prepared by geologic survey organizations;
- Geologic maps, especially old ones that depict the geology and landscape conditions prior to significant development;
- Geology texts and reference books;
- Interviews of local utility contractors or municipal maintenance departments;
- History of the geological formation along the pipe alignment; and
- Geotechnical reports from previous projects in the area.

Having a preliminary understanding of the structural properties of the rock mass prior to the advancement of subsurface explorations can allow the geotechnical engineer to undertake a targeted exploration drilling program that enables the location, terminal depth, and batter angle of the drill hole to be optimized to collect the greatest amount of useful data.

Thus, the second phase may consist of test borings, supplemented with one or more observation wells and geophysical exploration. Samples should be obtained for laboratory testing. Explorations and tests in this phase assist the designer in confirming the assumed initial pipeline alignment route and depth. On completion of the second phase, it is possible to establish a working geotechnical model for the site and go back to the field when required to develop additional location-specific geotechnical data. Observation wells enable the collection of groundwater information, including the fluctuation of groundwater levels. Groundwater samples can also be analyzed for water quality, potential contaminants, and corrosiveness.

The third phase of the investigation serves not only as an opportunity to obtain new information and verify issues discovered during the second phase but also as a validity test of the geotechnical model. The geotechnical professional will question his or her assumptions and preconceptions about the site and conduct various what-if scenarios. Depending on the outcome of this work and the importance of specific conclusions relative to subsurface conditions, a more detailed investigation with additional borings can be carried out to resolve any uncertainties about the ground conditions.

Other issues critical to the success of a subsurface investigation program include good-quality exploration and sampling services and the continuous on-site monitoring of the work by a trained geotechnical professional representing the engineer and/or owner. Thorough observation of field activities must be provided so that the results of the field work can be accurately and correctly interpreted.

4.3.2 Subsurface Geotechnical Explorations

It is essential that a thorough subsurface investigation be carried out to identify the geologic conditions along the pipeline alignment. The anticipated geologic conditions are often a significant factor in determining the feasibility and analyzing the risks of the proposed HAB project. The information compiled from the desktop study and limited soil borings, sampling, and laboratory analysis guides the selection of pipeline alignment and depth. More detailed soil borings, sampling, and laboratory analysis can be conducted to validate assumptions and obtain more specific data as needed for project design, construction, and risk management. Empirical field tests for soils, such as the SPT and cone penetrometer testing, are conducted. Test pits, trenches, and large-diameter drilled holes are used to assess the presence of cobbles and boulders or the extent of various geologic features and constituent materials that cannot be adequately described by the small sample from a test boring. This step is usually followed by extensive laboratory testing of the collected samples for physical properties, environmental and chemical properties, long-term loading conditions, and corrosion properties.

Frequently used types of subsurface sampling include

- Auger sampling,
- Split spoon (for SPT),
- Thin-wall sampling (Shelby tube for cohesive soils),
- Vibratory coring of soils,
- Rock coring,
- Test pits or trenches (appropriate means to assess the size and quantity of cobble and boulders),
- Hand augering, and
- Air-track probes to determine top of rock.

4.3.3 Exploratory Borings

Boreholes can be dug into the soil and/or bedrock to collect data on subsurface properties and samples for identification and laboratory testing. An important decision is to determine how many test borings to drill, the spacing along the proposed alignment, and the terminal depth. At a minimum, test borings should be taken at all jacking and receiving shaft locations. Additional borings are often appropriate for drives longer than 300 ft (90 m). If surface access is limited, other exploration methods—such as horizontal exploratory drilling, surface mapping, and/or geophysical methods—should be used to supplement the boring information. Test borings should be drilled below the shaft inverts to identify changes in the strata below the shaft that could affect the construction and long-term support of the proposed pipeline. The additional boring depth is needed to collect geotechnical information for the shaft foundation. If the subgrade of the shaft is below the GWT, the information about the soil below grade can be used to prevent the boiling of the subgrade soil and to select/design/install the dewatering system. A distance of at least 15 ft (4.5 m) below the subgrade elevation of any proposed shaft ground support system is recommended. If soft and compressible soils are present at or near the invert, the engineer should determine how many of the borings should be extended into firm ground.

It is important to survey the collar elevation at the ground surface of the borehole, since a difference of only a few inches may determine whether there is a mixed-face (soil and rock) condition. The borehole should not be in the location of the proposed alignment but should be off center to avoid creating a potential pathway for water to percolate to the working face [at least a 10-ft (3-m) offset from the centerline of the proposed bore is recommended]. It is crucial to properly backfill the borehole with cementitious grout after the drilling phase is complete.

On some projects, the designer may need to modify the auger boring invert during the design phase to avoid specific ground conditions. Extending the test boring depth to 15 ft below the shaft subgrade provides the designer the opportunity to select a more ideal alignment. Consequently, if the pipe invert is lowered, new borings (along the proposed alignment) should be drilled to the modified invert plus at least one pipe diameter or 5 ft (1.5 m), whichever is deeper. It is recommended that a geotechnical engineering professional make the final decisions on the number and depth of borings on the basis of the complexity of the project and the local ground conditions. The more difficult and costly the proposed project and the more complex the ground conditions, the more extensive the subsurface investigation program should be.

Important decisions that must be made about the test borings include the sampling interval and the need for laboratory and field testing. The frequency and types of samples that must be obtained are a function of the project type and ground conditions. Samples should typically be taken at intervals no greater than 5 ft (1.5 m) and at changes in strata to identify thin strata that may have significant effects on auger bore tool selection. In addition, continuous sampling from one diameter above the pipe to one diameter below the pipe is prudent. Disturbed sampling procedures can be used to classify the soil and rock. For laboratory strength and compressibility tests, more-costly undisturbed samples are required. Field tests that might be used are in situ vane testing, pocket penetrometer testing, point load testing on rock, pressure meter testing, and pump testing for groundwater hydraulic conductivity.

4.3.3.1 Soil Borings Soil borings are typically advanced using either a rotary-wash casing advance system or a hollow stem auger (HSA). Each drilling method is dependent on the anticipated ground conditions and subject to different cost constraints. Both methods allow for the insertion of internal sampling rods for SPTs. SPTs are one of the most widely used methods of sample collection and allow for the generation of a blow count number (N). A standard sampler is driven into the ground by a 140-lb hammer, dropped from a height of 30 in. (760 mm). The number of blows required to drive the sampler every 6 in. (150 mm) is counted and recorded. The first 6 in. of each sample in each interval is typically excluded because it is considered disturbed, but the number of blows for the second and third 6-in. drives is used as the N-value. The total drive lengths are typically 18–24 in. (455–610 mm); however, continuous sampling is also possible. The blow count number (subject to certain corrections related to sampling/drilling equipment and geostatic stress state) is frequently used for geotechnical engineering correlations to in situ soil properties and is known as

Table 4-1. Soil Classification Based on SPT N-Value

Sand and Gravel (Cohesionless Soil)		Clay and Silt (Cohesive Soils)	
SPT N-value	Relative Density	SPT N-value	Consistency
0–4	Very loose	0–2	Very soft
4–10	Loose	2–4	Soft
10–30	Medium	4–8	Medium stiff
30–50	Dense	8–15	Stiff
>50	Very dense	15–30	Very stiff
—	—	>30	Hard

Source: Florida DOT (2004).

the SPT N-value. The SPT is a relatively simple test but a useful index of soil characteristics. If an auto-hammer (usually at N85 to 95) is used, calibration (through an equation) is needed to convert to N60. Table 4-1 provides soil classifications based on the N-value.

4.3.3.2 Rock Coring Rock coring (or core borings) can also be advanced through casings in overburdened soils and seated down to the bedrock. Such methods of casing advance include conventional wash boring, casing hammers (if the overburden is shallow), and more recently, Symmetrix or ODEX-style down-the-hole (DTH) hammers affixed to a casing bit. Coring is frequently completed in 5-ft (1.5-m) run lengths and can be double-tube or triple-tube sampled, depending on the budget and sample disturbance requirements. Each core is generated through either a Christiansen-style (X) or wireline (Q) core retrieval system, where X-style coring requires the removal of all the rods for a sample return and Q-style coring allows for the retrieval of only the sampling interval. Coring bits can be appropriately sized to cut an NQ [1.9-in. (50-mm)] or HQ [2.5-in. (65-mm)] diameter core; however, larger and smaller core sizes are available, depending on the project demands. Each core is inspected on site by an engineering geologist and collected in a box for subsequent laboratory testing. Most HSAs will not accommodate rock coring because of casing size limitations.

Core holes can be vertical (which is frequently the case) or inclined, depending on access limitations and project-specific requirements. Furthermore, rock cores can be postoriented through the use of a downhole acoustic/optical televiewer (A/O TV) logging sonde or through conventional core orientation methods. An oriented rock core is helpful in understanding the three-dimensional attributes of discontinuities (e.g., joints or bedding planes) within the rock mass fabric. For example, with projects that need to advance through terrain that has faults/shears, groundwater, or unstable rock slabs, the orientation of significant fractures in the rock will be important for project viability.

Rock cores should be inspected for general character and for the presence of discontinuities. The core log must, at a minimum, include the time per foot of core drilled; the loss or gain of drill fluid or water; discontinuities during drilling, such as drill string drops; descriptions of the rock type and basic mineralogy; a weathering/alteration index; the field-estimated compressive strength; the core length recovered versus the length drilled; and the rock-quality designation (RQD). A more detailed discontinuity log should also be prepared and should include the attributes of the joints or bedding planes, such as the set spacing, the aperture (i.e., opening) width, the presence and type of infilling material, the joint condition, and the joint roughness.

The user is referred to industry-accepted standards of soil and bedrock drilling, sampling, and identification, as provided for in ASTM D2113, *Standard Practice for Rock Core Drilling and Sampling of Rock for Site Exploration*, and those referenced within these standards.

4.3.4 Test Pits/Trenches and Surface Mapping

At some sites, it is good practice to supplement the test boring program with other forms of subsurface investigation, such as test pits or large-diameter drill holes. Test pits and large-diameter drill holes provide the following:

- Large-volume samples,
- Exposure of undisturbed ground for observation,
- Direct evidence of the type and distribution of objects that exist in the ground, and
- Important facts about ground behavior and groundwater inflows.

For ground suspected to contain cobbles and boulders or other forms of obstructions, test pits and trenches can be an effective exploration technique. Test pits and trenches can be excavated by hand or with a backhoe to provide an accurate assessment of the size and likely frequency of boulders and cobbles, the grain size distribution, and the nature and size of potential obstructions (URS Greiner Woodward Clyde 2000). OSHA requires that pits deeper than 5 ft be adequately supported or shored unless the excavation is sloped or made entirely in stable rock; local codes may be more stringent.

If adjacent or local surface excavations and natural slopes present themselves, it is wise to gather additional information on geologic materials. Such information from field outcrops can be correlated to the boring logs and can help build a three-dimensional picture of rock (and soil) distribution.

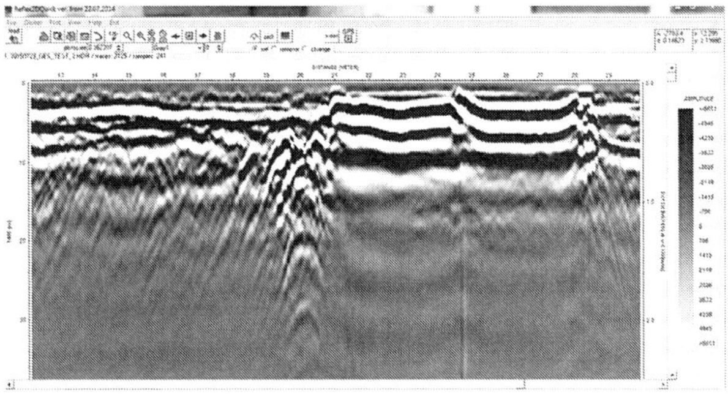

Fig. 4-1. Example image from GPR
Source: Geomatrix Earth Science Ltd. (2017)

4.3.5 Geophysical Investigation

Geophysical techniques and other emerging methods of nondestructive testing (NDT) may be effective in interpolating the subsurface conditions between the test borings, which are usually spaced along the bore alignment. These methods have been shown to be effective in some types of subsurface profiles. The extent of the investigation may vary depending on the local geological conditions and the risk management plan for the project. Subsurface investigation programs should identify general subsurface conditions and target the specific properties that could cause the project to fail, as identified by the initial desktop studies.

The geophysical subsurface investigation of soil deposits can be carried out by the following methods:

- Seismic refraction, which is good for shallow top-of-rock applications [typically <100 ft (30 m)];
- Seismic reflection, which is good for distinct units, faults, and increased depth penetration;
- Ground-penetrating radar (GPR) (Fig. 4-1), which is good for buried obstructions in soil;
- Cross-hole sonic logging;
- Electrical resistivity studies; and
- A/O TV logging.

4.4 GEOTECHNICAL PROPERTIES

Soil and rock characteristics that can significantly impact trenchless construction should be addressed in the geotechnical report. The following sections cover these characteristics and methods to obtain the required data. The relative importance of individual characteristics will vary from project to project, and no particular significance should be placed on the order in which the characteristics are listed and discussed. The following geotechnical properties should be considered during design for auger boring projects:

- Soil conditions;
- Potential for boulders, including their potential sizes and frequency;
- Rock conditions;
- Mixed-face conditions;
- Potential buried objects;
- Groundwater conditions;
- Potential for naturally occurring gas, such as methane and hydrogen sulfide;
- Contaminated groundwater or ground; and
- Artificial and environmentally sensitive features.

4.4.1 Soil Characteristics

It is important to understand the soil condition and potential reaction to the excavation process because soil condition will affect most, if not all, aspects of the boring process. Some ground can include coarse fraction (i.e., gravel and rock fragments) that significantly affects the ground behavior and boring performance. The soil type and properties impact the selection of the auger bore tooling, the production rate, and the ground stability. Laboratory testing to determine soil properties should be done in accordance with ASTM standards.

For cohesionless soils (predominantly sands and gravels), the important characteristics that should be identified include

- Grain size distribution,
- Coarse fraction,
- Cobbles and boulders (determined from test pits or large-diameter borings),
- Unit weight,
- Hydraulic conductivity, and
- Density (typically in terms of SPT N-values).

For cohesive soils (clays and silts), the important characteristics that should be identified include

- Moisture content,
- Plasticity index,
- Unit weight,
- Shear strength,
- Compressibility,
- Consistency (typically in terms of SPT N-values),
- Grain size distribution and hydrometer analysis,

- Coarse fraction,
- Stickiness/adhesiveness, and
- Swollenness.

Tunnelman's Ground Classification system, first described by Terzaghi (1950) and slightly refined by Heuer (1974), is designed to describe ground behaviors resulting from the excavation process and their impact on larger, conventionally constructed soft ground tunnels. The system is also a useful tool for the evaluation of soft ground conditions as they relate to auger boring. Descriptions of various ground behaviors, in terms of the Tunnelman system, and their impacts on boring are presented elsewhere (Bennett et al. 1995). General ground behaviors include:

- Raveling,
- Running,
- Flowing,
- Squeezing, and
- Swelling.

Table 4-2 provides a general description of different ground behaviors as they relate to stand-up times for conventional tunneling methods.

The clogging potential, also known as the stickiness/adhesiveness potential, is the tendency of plastic clay to adhere to the cutterhead and the augers, reducing productivity and/or halting production. Therefore, to prevent the clay from sticking to the cutterhead, the augers, and the outside surface of the casing, additional measures are needed if sticky clays are expected. The clogging potential is evaluated on the basis of the material's in situ moisture content and does not consider the effect of lubricants or other fluids introduced into the subsurface that could change the moisture content of the soil.

If clay soils are expected, it is helpful to measure the level of adhesion or stickiness to the cutterhead, the augers, and the outside surface of the casing. The plasticity index, the liquid limit, the water content, and the chemical composition and structure are the factors that affect the clay adhesion. The plastic limit, the liquid limit, and the moisture content are usually measured during the geotechnical investigation. Fig. 4-2 presents the low, medium, and high zones of clay adhesiveness. If the geotechnical engineer calculates the plasticity index (PI = liquid limit – plastic limit) and the consistency index [Eq. (4-1)] of the clay and plots them as shown in Fig. 4-2, he or she can make an informed recommendation of the levels of adhesion/stickiness.

$$\text{consistency index} = \frac{\text{liquid limit} - \text{water content}}{\text{liquid limit} - \text{plastic limit}} \tag{4-1}$$

The potential for clay soils to swell is a critical factor in the selection of the type of lubricant. Some clay soils swell when in contact with water,

Table 4-2. Tunnelman's Ground Classification for Soils

Classification		Behavior	Typical Soil Types
Firm		Heading can be advanced without initial support. Final lining can be constructed before ground starts to move.	Loess above the water table; hard clay, marl, cemented sand, and gravel when not highly overstressed.
Raveling	Slow raveling Fast raveling	Chunks or flakes of material begin to drop out of the arch or walls sometime after the ground has been exposed. This behavior is due to loosening or to overstress and brittle fracture (ground separates or breaks along distinct surfaces, as opposed to squeezing ground). In fast-raveling ground, the process starts within a few minutes; otherwise, the ground is slow raveling.	Residual soils or sand with small amounts of binder may be fast raveling below the water table and slow raveling above. Stiff fissured clays may be slow or fast raveling, depending on the degree of overstress.
Squeezing		Ground squeezes or extrudes plastically into tunnel, without visible fracturing or loss of continuity and without a perceptible increase in water content. Ground has ductile, plastic yield and flow due to overstress.	Ground with low frictional strength. Rate of squeeze depends on degree of overstress. Occurs at shallow to medium depth in clay of very soft to medium consistency. Stiff to hard clay under high cover may move in combination with raveling at excavation surface and squeezing at depth behind surface.

(Continued)

Table 4.2. Tunnelman's Ground Classification for Soils *(Continued)*

Classification		Behavior	Typical Soil Types
Running	Cohesive running Running	Granular materials without cohesion are unstable at a slope greater than their angle of repose (from ±30 to 35 degrees). When exposed at steeper slopes, the materials run like granulated sugar or dune sand until the slope flattens to the angle of repose.	Clean, dry granular materials. Apparent cohesion in moist sand or weak cementation in any granular soil may allow the material to stand for a brief period of raveling before the material breaks down and runs. Such behavior is cohesive running.
Flowing		A mixture of soil and water flows into the tunnel like a viscous fluid. The material can enter the tunnel from the invert as well as from the face, crown, and walls and can flow for great distances, completely filling the tunnel, in some cases.	Below the water table in silt, sand, or gravel without enough clay content to give significant cohesion and plasticity. May also occur in highly sensitive clay when such material is disturbed.
Swelling		Ground absorbs water, increases in volume, and expands slowly into the tunnel.	Highly preconsolidated clay with a plasticity index in excess of about 30, generally containing significant percentages of montmorillonite clay.

Source: Heuer (1974).

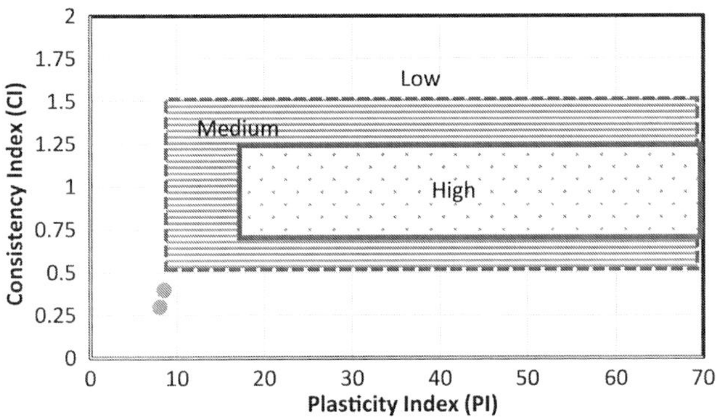

Fig. 4-2. Clogging potential of clay
Source: Thewes and Burger (2005)

reducing the overcut and increasing the friction between the soil and the casing, and consequently, the jacking force. Proper lubricant with hydration inhibitors reduces and prevents the swollenness and stickiness of clay.

4.4.2 Rock Conditions

On the basis of appropriate descriptive criteria (i.e., from ASTM or ISRM), the field subsurface and surface site investigation strategy for projects in which rock is anticipated should determine the following:

- Depth and extent of bedrock,
- Rock type (e.g., lithology and classification),
- RQD,
- Weathering/alteration index,
- Discontinuity set frequency/spacing,
- Discontinuity surface characteristics (e.g., aperture, infill material, roughness, shape, and joint roughness coefficient),
- Presence of fracture-controlled groundwater,
- Discontinuity orientation (e.g., dip and dip direction), and
- Estimated intact rock UCS (via field index and point load strength).

Conventional rock coring accompanied by DTH A/O TV logging can be a cost-effective way to arrive at the field data outlined above. This coring process also allows the user to collect data on discontinuity orientation.

After the field investigation phase, laboratory testing should be completed on select samples obtained in the field to arrive at reasonable estimates of the following:

- Hardness (e.g., Shore),
- UCS,

- Direct shear strength,
- Tensile strength (e.g., Brazilian),
- Punch penetration,
- Abrasiveness (e.g., Cerchar and Taber),
- Mineralogical (thin sections), and
- Slake durability and swelling tendencies (for shale).

Depending on the depth of burial, the geologic region of the proposed work, and the project risk, additional field testing may be necessary and may include the determination of the in situ rock mass modulus of deformation or horizontal stresses through pressure meter testing. In addition, packer testing (e.g., the Houlsby method) can be very useful in determining fracture-controlled hydraulic conductivity and the behavior of discontinuities under controlled head conditions.

Reliable exploration and test methods are available for determining the rock mass characteristics listed above. Drilling and sampling procedures that both maximize core recovery and minimize core damage are essential to the development of a reliable rock mass characterization program. Depending on the quality of the rock, in less massive rock formations triple-tube (e.g., HQ and NQ) coring generally provides the least sample disturbance. When collecting rock cores, it is wise to take photographs with a scale and color guide immediately after opening the sample. Once the sample is opened or moved, it will begin to deviate from its in situ condition because of exposure to the atmosphere and vibration.

Developed by Deere (1963), RQD is widely used as an indicator of overall rock mass quality and was the first rock mass classification method. RQD is one of the most widely used and referenced indexes for construction in bedrock and is calculated as the sum of the lengths of the intact cores greater than 4 in. (100 mm), divided by the total core run length (expressed as a percentage). Rock quality determinations can be incorporated within empirical rock mass classification schemes (e.g., the rock mass rating (RMR) and Q) that were developed for the construction of larger tunnels (Bieniawski 1974; Barton et al. 1974). Rock mass classifications can be used to develop estimates for rock mass properties and behavior, like the modulus of deformation and the stand-up time, respectively. RQD and the subsequent rock mass classification schemes can vary with respect to orientation. When drilling only vertical holes, engineers must be aware that the recorded RQD may not be the lowest RQD, unless it is based on the volumetric joint count method presented by Palmstrom (2005).

The weathering index, the fracture frequency, the joint characteristics, and the overall rock mass quality are also important for estimating borability (the capacity of the rock in front of the cutterhead to be ripped, chipped, and broken into smaller pieces to allow the advance of the cutterhead). The estimation of block sizes can be an important factor in evaluating the potential for blocks to become wedged in the cutterwheel or to become separated or dislodged above the pipe and gouge the pipe or couplings.

Abrasiveness and unconfined compressive and tensile strength are key parameters in the estimation of borability. Point load tests and Schmidt hammer tests can be used to supplement laboratory-determined UCS data and rapidly obtain preliminary strength data in the field. The boundary between soil and rock UCS is commonly described as 140 psi [20,000 psf (965 kPa)]; however, at such low rock strengths, the boundary is subject to interpretation on the basis of site-specific bedrock properties.

Bedrock mineralogy impacts the efficiency of the cutterhead at the advancing face. The preparation of thin sections and evaluation by a qualified geologist will yield important details on sample mineralogical composition at the intact rock scale. For example, a high percentage of siliceous mineral content (e.g., quartz) will typically result in increased abrasiveness of the rock, which may result in accelerated rates of bit or cutter wear. The same potentially problematic mineralogy at the advancing face may also pose problems of increased base friction (e.g., increased jacking/thrust forces) parallel to the drive length, which may impact the need for lubricating fluids.

Borability is controlled by the frequency of discontinuities of the rock, best determined on the basis of RQD (ASTM D6032), as well as by laboratory-estimated intact rock properties, such as unconfined (uniaxial) compressive strength (ASTM D7012), tensile strength (ASTM D3967), hardness (ASTM D5873), and the abrasiveness of the intact rock (ASTM D7625). The borability of the rock also affects the tooling efficiency and the tooling survivability. Borability typically falls into four categories: no impact to tooling, minor impact to tooling, difficult, and rock dominance. Some rock may be too tough, unfractured, and abrasive to be economically excavated with typical tooling.

4.4.3 Mixed-Face Conditions

Mixed-face conditions are defined as distinct variations in ground conditions within the cross-sectional area of the face, such as soil over rock. Mixed-face conditions should be identified, and the vertical and/or horizontal alignments should be modified during the design phase to avoid these conditions, to the extent possible. Identification may require the use of extremely close-spaced borings or probes to locate the mixed face. In addition, geophysical surveys complemented by borings can provide indications of the top of the rock and hard layers. Mixed-face conditions can result in uneven forces and wear on the cutterhead, thereby reducing steering control and stability of the face. These conditions also present the risk of overexcavation and the potential to halt the machine's advance.

If mixed-face conditions cannot be avoided, they should be documented in the contract documents, which should require bidders/contractors to assess the risk and develop strategies to address the mixed-face conditions. These potential problems can be partially or completely mitigated through the proper design of the cutterwheel, the use of ground stabilization, and the careful operation of the cutterhead.

4.4.4 Natural Obstructions: Gravels, Cobbles, and Boulders

The extent, frequency, size distribution, and physical properties of gravels, cobbles, and boulders should be determined. The pipe alignment should be located to minimize the amount of gravels and cobbles to be encountered, and to the extent possible, avoid boulders. If these conditions cannot be avoided, the auger bore should be operated and selected with features designed to reduce risk when excavating through cobble- and boulder-rich ground. If the anticipated size and frequency of cobbles and boulders will obstruct the cutterhead or overwhelm the auger spoils removal system, an allowance for obstruction removal or cutterhead rescue should be provided in the bid documents.

The presence of gravels, cobbles, and boulders is a function of the geological setting, including glacial deposits, moraines, tills, lag zones, alluvial deposits, colluvial or talus deposits, and weathered rock (core stones). The gravels, cobbles, and boulders can be isolated or nested, depending on their position within the deposits.

Intensely packed gravels or the presence of cobbles and/or boulders encountered by an auger bore can bring an auger boring project to an abrupt halt. If rock fragments are encountered in the sampler during the geotechnical investigations, steps should be taken to determine whether the rock pieces are naturally occurring gravel or were generated by the sampler encountering and fracturing cobbles and/or boulders. When there is a potential for gravels, cobbles, and boulders, large-diameter boreholes [up to 3 ft (1 m)], rotosonic borings, test pits, or trenches can be useful in obtaining representative samples for grain size analysis and determining the size and distribution of the gravels, cobbles, and boulders. Samples of the matrix material should also be obtained and tested.

The unknown orientation of the odd-shaped cobble or boulder is important. The best method of sizing cobbles and boulders is based on determining the largest dimension of the rock in any direction. The physical properties of the cobbles and boulders should be determined. The rock type and mineralogy of the gravels, cobbles, and boulders should also be determined to help in selecting the best equipment for the project.

One way to document the frequency of the cobbles and boulders is to determine the cobble volume ratio (CVR) and the boulder volume ratio (BVR). The CVR and BVR are the volumes of cobbles and boulders, respectively, in relation to the excavation volume, reported in terms of percentage. CVR and BVR are best determined by digging test pits or trenches with a calculable volume. For individual isolated cobbles or boulders in a test pit, CVR and BVR can be as small as 0.01%. A zone of nested cobbles or boulders in a test pit can have a CVR and BVR as high as 35%. The CVR and BVR values can be checked and confirmed when the entry and exit pits are excavated.

4.4.5 Potential Obstructions

Establishing the likelihood that buried objects will be encountered requires an evaluation of information from a variety of sources, including regional and site geology reports, site usage history, geophysical surveys, borings, and test pits. Field investigations—including test borings, test pits, and under some circumstances, geophysical surveys—can serve to verify preliminary conclusions about the potential for encountering obstructions.

Borings and test pits provide physical evidence of the nature, location, frequency, and size of the buried objects that may be encountered. Geophysical surveys can sometimes provide insight into the relative sizes of buried objects, the frequency of their occurrence, and the locations of anomalies, depending on the depth of interest and resolution. Indications of anomalies, based on geophysical surveys and other information, should be verified with borings and test pits.

Although there are many approaches for locating potential obstructions, locating all of them along a given alignment is often not possible within practical budget and time constraints for site investigations. This situation results in the need to develop plans for dealing with obstructions, technically and contractually, before construction begins.

4.4.6 Groundwater Conditions

The presence of water in a boring operation has the potential to be disastrous. A preliminary survey should be conducted on the site to study the existing water table conditions before boring and proper dewatering techniques are implemented. Groundwater conditions will have a significant influence on ground behavior, and therefore on the design and construction of the jacking and receiving shafts and the selection and operation of the auger boring tooling.

Water table conditions may change during the year. The groundwater level and piezometric levels in confined aquifers should be determined by installing observation wells and/or piezometers at shaft locations. Consider adding piezometers at intermediate points for longer drives. Piezometers installed at shaft locations during design can later be used during shaft excavation to confirm shaft watertightness.

The hydraulic conductivity of water-bearing strata should also be determined. The risk of not successfully completing an auger boring operation becomes elevated as the groundwater reaches about a third of the height of the cutterhead, and higher water levels can require the use of extensive dewatering systems to allow the auger bore to safely proceed.

Dewatering can increase friction on the pipe and result in higher jacking forces because of increases in the effective stress and the loss of lubrication. These dewatering impacts should also be considered in the selection of shaft excavation support, since some systems require dewatering in permeable soils.

4.4.7 Contaminated Ground or Groundwater

The potential for encountering contaminants should be determined during the planning phase. If contaminants are found, then determination of their nature and extent must be undertaken during the site investigation. Even if contaminants are not identified during the planning phase, the site investigation should still look for and test for contaminants.

As with any type of construction, unanticipated ground and/or groundwater contamination will impact project costs and schedules. Careful planning and execution may minimize these impacts. Even if ground or groundwater contamination is anticipated, its presence can dramatically increase the cost of a project, depending on the types of contaminants and their extent. However, auger boring can be very advantageous when contaminated materials are encountered, because the quantities of contaminated ground and groundwater that must be handled are significantly less than would be handled by open-cut methods. This statement assumes that the materials do not pose a hazard to the labor force.

Sources of information on the potential for contamination include health department records; fire, planning, and building departments; and the local and regional agencies responsible for pollution control and water quality. In areas where "paper studies" indicate the potential for contamination, chemical analyses of ground and groundwater samples should be undertaken to identify the contaminants' nature and concentration. This information is necessary to determine the appropriate procedures for disposing of ground and groundwater during construction.

Personnel training for the handling of contaminated ground and groundwater should be implemented and followed per local, state, and federal requirements during all phases, including site investigation and construction.

4.4.8 Artificial and Environmentally Sensitive Features

Existing site features that could affect shaft construction and auger boring operations should be identified during the site investigation. Existing utilities may be located by consulting as-built records or contacting locating services, although the locations of abandoned utilities are often not known. Potholing should be used to confirm locations of utilities that are near or intersected by the planned pipeline. Historic buildings and environmentally sensitive areas usually require evaluation on a case-by-case basis to ensure their protection. Archaeological significance may apply to specific findings or historic sites.

One of the most problematic obstacles is buried footings and the remnants of foundations left behind in the ground. The demolition of artificial structures typically involves removing all aboveground portions (the visible parts) and then cutting the top few feet of the foundations below the ground surface. The remaining structures are potential objects for the auger bore to encounter. It is important to investigate the history of the proposed alignment and look for structures that were once in place. A review of historical photographs over set periods of time is one way to identify potential artificial and buried objects.

4.4.9 Laboratory Testing

The typical laboratory tests recommended for cohesive soils (silts and clays) are moisture content and unit weight (ASTM D2216 and D2937), hydrometer analyses of fine fraction (ASTM D422), Atterberg limits (ASTM D4318), specific gravity (ASTM D854), and UCS (ASTM D2166). For cohesionless soils (sands and gravels), the recommended tests are moisture content and unit weight (ASTM D2216 and D2937), grain size analyses (ASTM D422), specific gravity (ASTM D854), and permeability tests (ASTM D2434). For rock, the recommended tests are moisture content and unit weight (ASTM D2216 and D2937) and UCS (ASTM D2938).

4.5 GEOTECHNICAL REPORTING REQUIREMENTS

Once surface, subsurface, and laboratory data are collected, reviewed, and finalized, a deliverable geotechnical report is generated. Such reports can take the form of a conventional stand-alone comprehensive document that includes results from the field investigation program and design recommendations.

A geotechnical data report (GDR) can be generated, as suggested in Essex (1997), which is a joint publication of the Underground Technology Research Council (UTRC), ASCE, and the American Institute of Mining, Metallurgical and Petroleum Engineers. The GDR is not considered an interpretive document; however, it is never entirely free of some level of interpretation, resulting from field practices. Interpretation as it relates to use of such field data for design purposes should be avoided in the GDR. The GDR should contain a summary of hard data, which are collected and measured in the field or in the laboratory, and should be included in the contract documents as a stand-alone volume or provided on contract documents, depending on the size of the project and the amount of information. The selection of appropriate geotechnical reporting documents may vary by project, depending on the contracting mechanism (e.g., design–build ventures may have different reporting requirements than traditional design–bid–build projects).

Often, it is necessary to produce a series of GDMs or a geotechnical interpretive report (GIR). These memoranda or reports typically present summaries of the geotechnical data, interpretations of the data, information about earth pressures to be used for design, discussions of the expected behavior of the ground, and other geotechnical information, as well as design recommendations, such as appropriate boring and shaft types and systems. Since these reports and memoranda are typically prepared prior to design, much project information is included that may not be applicable at the time of the bid. In addition, there are typically many broad statements about what may or may not occur during construction. Although such reports are useful to the designer, they can be misleading or even erroneous at the time of the bid and should not be part of the contract documents. It is standard practice to disclaim these reports and supersede them with more appropriate language and baseline statements in the contract.

Increasingly, another stand-alone report is being prepared, known as a geotechnical baseline report (GBR). The GBR is typically prepared with the owners' input as the design is being completed and serves as the definitive geotechnical baseline for use in the resolution of disagreements, disputes, or claims related to differing subsurface conditions. The GBR, prepared by the engineer, must be made available to the contractor before bidding for the project. This report is described in great detail in *Geotechnical Baseline Reports for Construction: Suggested Guidelines* (Essex 2007). Readers are strongly urged to refer to this publication for more detailed information on GBR history, rationale, and preparation. The contractor is responsible for risks associated with conditions that are consistent with, or less adverse than, those indicated in the baseline report. However, if the subsurface site conditions are significantly more adverse than those reported in the baseline, the owners should pay the additional cost of overcoming those conditions. The report should include the following information at a minimum:

- Project description,
- Available existing information/reports,
- Description of the geology,
- Generalized geologic profile,
- Groundwater conditions,
- Presence of contaminants,

Table 4-3. Crucial Minimum Subsurface Parameters Needed by Contractors

Soil Parameters	Rock Parameters
Identification	Identification
Gradation	Color
Permeability	Mineral grain size
Density (or unit weight)	Composition
Standard penetration value (blow count)	Intact rock strength (UCS)
Cohesion	Hardness & abrasiveness
Moisture content	Quartz content
Atterberg limits	Fracture frequency
UCS	RQD
Groundwater	Total core recovery

- Geological features that may impact the project,
- Artificial features that may impact the project,
- Anticipated ground behavior at the shaft locations, and
- Anticipated ground behavior along the pipeline.

Table 4-3 presents the subsurface parameters that are crucial and have to be supplied to the contractor for preliminary assessment.

Both the GDR and GBR become contract documents and, as such, are part of what the contractor will rely on to develop a project construction cost and approach. The GDM or GIR is an intermediate report intended solely for the owners' benefit and is not normally part of the contract documents.

Completeness of the boring information is essential and should include the following information (see Fig. 4-3):

- Ground surface elevations,
- Methods used to obtain soil samples,
- Standard penetration resistance values (blow counts or N-values),
- Detailed soil descriptions with accurate depiction of soil layers and transitions,
- Soil strength values and how they are determined,
- Moisture content and dry density of samples,
- Water-level data (encountered depth, at completion, prior to back-filling, and long-term levels are all important),
- Method used to backfill and seal hole, and
- Gas conditions.

Geotechnical investigations and the reports that follow should also focus on conditions that may make HAB unfeasible. Some of the most problematic ground conditions for the HAB method include

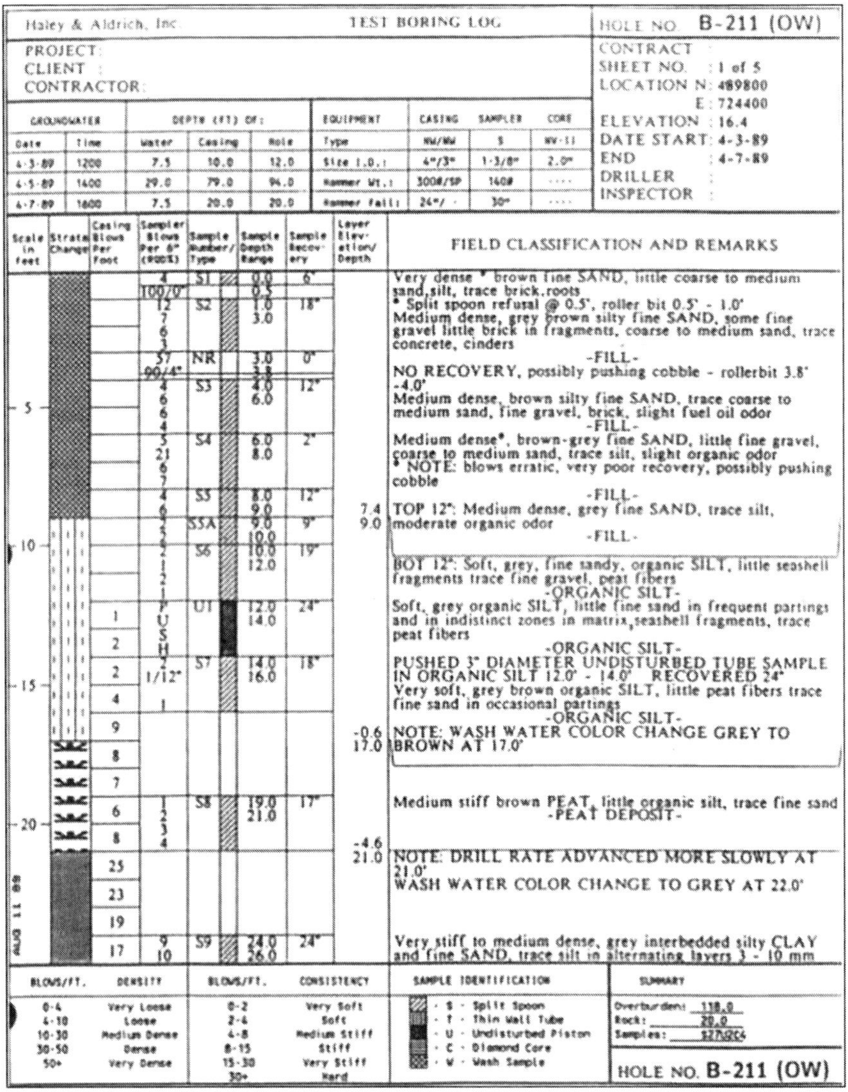

Haley & Aldrich, Inc. TEST BORING LOG HOLE NO. B-211 (OW)

PROJECT:
CLIENT:
CONTRACTOR:

CONTRACT:
SHEET NO.: 1 of 5
LOCATION N: 489800
E: 724400
ELEVATION: 16.4
DATE START: 4-3-89
END: 4-7-89
DRILLER:
INSPECTOR:

GROUNDWATER		DEPTH (FT) OF:		
Date	Time	Water	Casing	Hole
4-3-89	1200	7.5	10.0	12.0
4-5-89	1400	29.0	79.0	94.0
4-7-89	1600	7.5	20.0	20.0

EQUIPMENT	CASING	SAMPLER	CORE
Type	NW/NW	S	NV-II
Size I.D.:	4"/3"	1-3/8"	2.0"
Hammer Wt.:	300#/SP	140#	....
Hammer Fall:	24"/ -	30"	....

Sample Number/Type	Sample Depth Range	Sample Recovery
S1	0.0–0.5	6"
S2	1.0–3.0	18"
NR	3.0–3.8	0"
S3	4.0–6.0	12"
S4	6.0–8.0	2"
S5	8.0–9.0	12"
S5A	9.0–10.0	9"
S6	10.0–12.0	19"
U1	12.0–14.0	24"
S7	14.0–16.0	18"
S8	19.0–21.0	17"
S9	24.0–26.0	24"

Sampler blows per 6" (RQD%): 4, 100/0"; 12, 7, 6, 3; 57, 90/4"; 4, 6, 6, 4; 5, 21, 6, 7; 4, 6; 4, 5; 6, 2, 1, 2, 1; PUSH; 1/12", 1; 1, 2, 3, 4; 9, 10

Casing blows per foot: 1, 2, 2, 4, 9, 8, 7, 6, 8, 25, 23, 19, 17

Layer elevation/depth: 7.4 / 9.0; -0.6 / 17.0; -4.6 / 21.0

FIELD CLASSIFICATION AND REMARKS

Very dense * brown fine SAND, little coarse to medium sand, silt, trace brick, roots
* Split spoon refusal @ 0.5', roller bit 0.5' - 1.0'
Medium dense, grey brown silty fine SAND, some fine gravel little brick in fragments, coarse to medium sand, trace concrete, cinders
-FILL-
NO RECOVERY, possibly pushing cobble - rollerbit 3.8' -4.0'
Medium dense, brown silty fine SAND, trace coarse to medium sand, fine gravel, brick, slight fuel oil odor
-FILL-
Medium dense*, brown-grey fine SAND, little fine gravel, coarse to medium sand, trace silt, slight organic odor
* NOTE: blows erratic, very poor recovery, possibly pushing cobble
-FILL-
TOP 12": Medium dense, grey fine SAND, trace silt, moderate organic odor
-FILL-
BOT 12": Soft, grey, fine sandy, organic SILT, little seashell fragments trace fine gravel, peat fibers
-ORGANIC SILT-
Soft, grey organic SILT, little fine sand in frequent partings and in indistinct zones in matrix, seashell fragments, trace peat fibers
-ORGANIC SILT-
PUSHED 3" DIAMETER UNDISTURBED TUBE SAMPLE IN ORGANIC SILT 12.0' - 14.0' RECOVERED 24"
Very soft, grey brown organic SILT, little peat fibers trace fine sand in occasional partings
-ORGANIC SILT-
NOTE: WASH WATER COLOR CHANGE GREY TO BROWN AT 17.0'
Medium stiff brown PEAT, little organic silt, trace fine sand
-PEAT DEPOSIT-
NOTE: DRILL RATE ADVANCED MORE SLOWLY AT 21.0'
WASH WATER COLOR CHANGE TO GREY AT 22.0'
Very stiff to medium dense, grey interbedded silty CLAY and fine SAND, trace silt in alternating layers 3 - 10 mm

BLOWS/FT.	DENSITY	BLOWS/FT.	CONSISTENCY
0-4	Very Loose	0-2	Very Soft
4-10	Loose	2-4	Soft
10-30	Medium Dense	4-8	Medium Stiff
30-50	Dense	8-15	Stiff
50+	Very Dense	15-30	Very Stiff
		30+	Hard

SAMPLE IDENTIFICATION
- S - Split Spoon
- T - Thin Wall Tube
- U - Undisturbed Piston
- C - Diamond Core
- W - Wash Sample

SUMMARY
Overburden: 118.0
Rock: 20.0
Samples: S27U2C4

HOLE NO. B-211 (OW)

Fig. 4-3. Example of a soil boring log
Source: Doherty (2016)

- Soils containing boulders. Geotechnical reports should comment on the potential for encountering nested cobbles and boulders, and their potential size and frequency.
- "Running" sand. Dry, loose to very loose, or medium to coarse sands that may enter into and envelop the excavation in an uncontrollable fashion.

- Saturated soil or bedrock conditions below the water table, including joint-controlled water inflow within bedrock.
- Variable bedrock conditions. For example, relatively hard bedrock with thick clay infillings, mixed-face conditions, or alternating intervals of weathered and competent bedrock.
- Contaminated ground conditions, which can create health and environmental concerns and/or corrosive impacts on equipment.

REFERENCES

Alexander, C., and Marshall, M. I. (2006). "The risk matrix: Illustrating the importance of risk management strategies." <https://joe.org/joe/2006april/tt1.php> (Oct. 10, 2016).

Barton, N., Lien, R., and Lunde, J. (1974). "Engineering classification of rock masses for the design of tunnel support." *Rock Mech. Rock Eng.*, 6(4), 189–236.

Bennett, R. D., Guice, L., Khan, S., and Staheli, K. (1995). "Guidelines for trenchless technology: Cured-in-place pipe (CIPP), fold-and-formed pipe (FFP), mini-horizontal directional drilling (mini-HDD), and microtunneling." Rep. CPAR-GL-95-2 (Trenchless Technology Center Rep. 400), U.S. Army Corps of Engineers, Washington, DC.

Bieniawski, Z. (1974). "Geomechanics classification of rock masses and its application." *Proc., 3rd Int. Congress on Rock Mechanics*, ISRM, Denver, 27–32.

Deere, D. (1963). "Technical description of rock cores for engineering purposes." *Rock Mech. Eng. Geol.*, 1(1), 16–22.

Doherty, D. (2016). *Soil boring log*, Haley and Aldrich, Inc., Bedford, NH.

Essex, R. J., ed. (1997). *Geotechnical baseline reports of underground construction: Guidelines and practices*, ASCE, New York.

Essex, R. J., ed. (2007). *Geotechnical baseline reports for construction: Suggested guidelines*, ASCE, Reston, VA.

Florida DOT. (2004). "Soils and foundations handbook." State Materials Office, Gainesville, FL.

Geomatrix Earth Science Ltd. (2017). https://www.geomatrix.co.uk/software/ground-penetrating-radar/ (Jan. 17, 2017).

Heuer, R. E. (1974). "Important ground parameters in soft ground tunneling." *Subsurface Exploration for Underground Excavation and Heavy Construction*, ASCE, New York, 41–55.

Palmstrom, A. (2005). "Measurements of and correlations between block size and rock quality designation (RQD)." *Tunnels Underground Space Technol.*, 20(4), 362–377.

Peck, K. T. (1967). *Soil mechanics in engineering practice*, Wiley, New York.

Terzaghi, K. (1950). "Geologic aspects of soft ground tunneling." *Applied sedimentation*, Trask, P., ed., Wiley, New York.

Thewes, M., and Burger, W. (2005). "Clogging of TBM drives in clay—Identification and mitigation of risks." *Underground space use: Analysis of the past and lessons for the future*, Taylor & Francis Group, London, 737–742.

URS Greiner Woodward Clyde. (2000). "Microtunneling design guidelines." Pipeline Research Council International, Inc., Houston.

CHAPTER 5

DESIGN AND PRECONSTRUCTION PHASE

Design is a process of planning and making informed decisions to manage risks for a project that is being developed or built. This process requires knowledge of the materials, equipment, and construction procedures that culminate in plans and specifications that are based on sound scientific facts, principles, and experience. Accurate and detailed design documents permit a qualified contractor to accurately estimate the cost, properly allocate risks, and complete the project on budget and on schedule. Final design of trenchless projects occurs during construction. The final result is the as-built document.

Informed decisions made during the design process require facts and knowledge-based assumptions. Analyses of such facts and informed assumptions, along with experience, form an opinion on the applicability of HAB for a specific site. Initial assessment by the design team early on in the design phase of a project is used to identify hazards such as

- Utilities;
- Subsurface conditions;
- Social impacts, such as to traffic, businesses, and pedestrians;
- Hazardous substances;
- Wetlands and floods; and
- Seismic impacts.

Risk assessments for trenchless projects are used typically to determine the probability of a negative event or hazard occurring. HAB involves risks associated with subsurface hazards that can have significant adverse impacts on project safety, the environment, cost, and schedule. The management of these risks with all stakeholders from the initiation of a project will often determine overall project success. For example, an auger bore can encounter unanticipated bedrock or large boulders that may require 3 to 10 times the schedule and cost of an auger bore in soil. Some risks, should they occur, can result in expensive delays and project failure. As stated earlier, HAB can vary from simple short crossings in relatively uniform soils to complex crossings with variable ground conditions in critical locations with significant economic and/or environmental impacts. Therefore, before the start of the design phase, the engineer and owner make decisions regarding the depth and breadth of the design studies, and consequently, the scope and budget of the design work. In these decisions, risk trade-offs are made between identifying/studying some risks during the design phase and dealing with the impacts of these risks during the construction phase, if they occur. Key factors in the trade-off decisions are the knowledge and experience of the owner, engineer, and contractor of the specific site conditions; the limitations and abilities of the construction techniques; the probability of the risk; the cost of studying the risk during design and the potential risk reduction during construction; and the cost of the impact if the risk materializes.

5.1 FEASIBILITY AND RISK ASSESSMENT

During the planning phase, the engineer should consider potential project risks and ways to mitigate them to acceptable levels. This exercise helps evaluate the feasibility of auger boring for the proposed installation and identifies project-specific concerns that need to be addressed during design. Areas of potential risk associated with auger boring can include any of the following:

- Unanticipated subsurface conditions (either along the HAB alignment or within the shaft),
- Impacts of dewatering on adjacent structures,
- Damage to adjacent utilities,
- Surface heave or settlement,
- Line and grade problems,
- Pipe damage,
- Obstructions, and
- Work space limitations.

The identification and reduction of risk during project planning and design are effective means of risk management, which involves experts who are responsible for the following:

- Specifying performance criteria for the contractor's means and methods;
- Project funding and schedule;

- Environmental and social constraints;
- Subsurface conditions and the impact of subsurface conditions on design, construction means and methods, and mitigation options should the risks be realized;
- Permit conditions that may impact access, work space, and alignment selection; and
- Contingency planning.

Throughout the planning and design phases of the project, many assumptions are made about pipe location, soil conditions, alignment, and so on. In addition, risk management assumptions and decisions are made. Throughout the geotechnical exploration and design phases, these assumptions and risk decisions need to be checked and revalidated in light of the acquired knowledge.

5.2 AUGER BORING DESIGN

The auger boring design process typically consists of two components: long-term design and construction design. For conventional design–bid–build projects, the owners' design engineer is responsible for the long-term design and the performance specification for the contractor's means and methods. Long-term design elements are typically applied to long lead items and long-term performance requirements. The following project elements are typically considered as part of the long-term design:

- Project location;
- Historic site use;
- Pipe depth, size, grade, and pressure capacity (as applicable);
- General construction process (e.g., trenching versus trenchless);
- Specialty construction process if required by site conditions and civil restrictions (e.g., auger boring versus microtunneling);
- Easements for permanent and temporary construction needs;
- Utility identification and verified locations within 25 ft (7.6 m) of the excavation;
- Subsurface conditions and geotechnical properties for long-term and construction loads;
- Permit filing and appropriate conditions, such as minimum casing size, filling, and wall thickness, as required;
- Carrier pipe design to meet specified owner operation and performance criteria;
- Carrier and casing pipe design to meet project live and dead loads, including loadings that result from annular fill placed between the casing and carrier pipe;

- Determination of sensitive structures and design of minimum performance or support system to protect adjacent properties;
- Corrosion protection; and
- Minimum requirements for specialty services, such as dewatering and the handling of environmental issues or special project restrictions, such as may occur in urban construction.

Construction design is typically applied to ensure that certain construction elements are capable of withstanding the loads expected during the construction stage; these loadings are based on the construction equipment and OSHA requirements. Prior to construction, the contractor typically submits the design elements for review by the design team to ensure compliance with the design requirements and as a check on the anticipated product loading from construction.

The contractor is typically responsible for selecting and designing the following project elements, according to the data and performance criteria provided by the owners in the bid documents:

- Shoring/sloping the jacking and receiving pit,
- Dewatering system,
- Jacking equipment,
- Thrust blocks or other reaction system necessary for the jacking equipment,
- Casing for resisting jacking forces and potential misalignments,
- Spoils management and disposal conditions, and
- Nonstructural annular fill design.

The contractor is required to estimate the jacking forces to verify that the HAB machine can complete the bore without damaging the casing. For the conditions represented in the contract documents, the contractor is also responsible for verifying constructability when the specified construction techniques and industry standard construction practices are used. In certain cases, the contractor may be required to provide supplemental subsurface data and utility-locating services, although these responsibilities must be specified in the scope of work in the contract. In such a case, the owner may consider an appropriate additional schedule and cost to obtain permits and rights of entry, assess the data, and wait for the project redesign if warranted. Contractors should not be required to obtain new subsurface data prior to bidding a project, as there is typically insufficient time for such data collection.

5.2.1 Casing Pipe Design

The auger boring process involves advancing a steel casing with internal augers, which may serve as the product pipe or temporary casing for the product pipe (or pipes). If the casing is used as the product pipe without replacement, then the pipe must meet the design guidelines of the carrier pipe. The casing pipe has to be designed to withstand the axial and normal loads; the axial loads are discussed in Chapter 6. The interaction between the soil and the pipe should be considered in designing the pipe to withstand the normal loads. If the steel casing compresses or deforms under the soil load, soil pressure may be relieved by the soil's internal arching action as the pipe deforms. Flexible pipes derive their ability to support loads from their inherent strength, plus the passive pressure of the soil attributable to the outward movement of the pipes against surrounding soil as they deflect. Moreover, the soil itself may carry some vertical loads by the arching action within the soil. As the ring deflects, it redistributes and relieves itself of some of the soil pressure (Atalah 1996).

Excessive deformations, soil displacement, and pipe wall buckling are some of the performance limits beyond which the conduit cannot serve the designed purpose; these performance limits should not be exceeded. Once the deformation limits or performance limits have been set, the objective is to determine the pipe wall thickness that will prevent the pipe from reaching this limit. The design follows four simple steps:

1. Predict external pressure,
2. Determine adequate pipe wall thickness,
3. Check ring deflection, and
4. Check handling factor.

The external pressure is calculated as described in Section 5.2.3 and consists of dead load, superimposed load, and live load.

Generally, the steel casing used for HAB is thick enough that corrosion protection coatings are not required. Coating the casing pipe is not recommended because the coating will be damaged during installation from the rotation of the augers on the inside and the soil from the outside. It is recommended, however, that a pipe with greater wall thickness be used to increase the longevity of the pipe, especially if corrosive conditions are present.

The long-term external loads are the expected loads during the service phase of the pipe. These loads are the pressure from the overburden soil and the live load from trucks, trains, or airplanes. In pressure pipes, the service pressure also needs to be checked. Depending on the depth of installation and the subsurface conditions present, the casing needs to withstand significant long-term external loads. Casings that do not provide sufficient resistance to loading may suffer premature damage, including buckling. As such, the casing should be designed to withstand service pressure stresses, overburden loads, and (where applicable) external static water loads. The casing design should also consider all long-term in-service loads, including static and transient pressures.

5.2.2 Carrier Pipe Design

The goal of the carrier pipe design is to determine the material properties and dimensions necessary for the carrier pipe to withstand long-term, owner-defined criteria. Carrier pipe failure may be a result of unconstrained buckling from lateral loading and/or of thrust buckling from total vertical loading, joint failure, and so on. Casing failure can also occur from long-term corrosion, causing a reduction in the casing stiffness and strength. This failure is often addressed by filling the casing to the carrier annulus with a cementitious grout to eliminate the void and the subsequent surface settlement that can occur as a result of casing failure.

Successful design involves the determination of the loads on the pipe and the assessment of the loading conditions on the mechanical stability of the pipe. The pipe's capacity to withstand the applied loads should be higher than the applied load combinations by a predetermined factor of safety. Load sources include both construction (addressed in Chapter 6) and long-term conditions. Sources of these loadings include

- Soil and groundwater vertical loads;
- Live loads and impact loads from traffic on roadways and railroads and at airports;
- Surcharge loads;
- Thermally induced loads;
- Environmental loads during usage, such as flooding buoyancy loadings;
- Internal pressure;
- Load reduction over time due to corrosion; and
- Other owner design criteria.

After the successful installation of the casing pipe, the carrier pipe (also referred to as the product or distribution pipe) can be installed. This installation is accomplished by either using the carrier pipe to push the (temporary) steel casing into the exit pit or by installing the carrier pipe within the (permanent) steel casing. The carrier pipe is designed to perform the hydraulic function of the pipe in terms of flow volume, pressure, velocity, and so on. This pipe also must withstand the expected construction and service loads. The pipe's hydraulic design and material selection are outside the scope of this manual. The expected loads during the construction and service phases of the pipe are presented in the next paragraphs.

If the carrier pipe is used to push (extrude) the steel casing out, the carrier pipe needs to be designed to withstand the jacking forces associated with this operation. The short-term design considerations for such a pipe are similar to those for the casing. For a carrier pipe installed within the casing (permanent casing), wooden skids or premanufactured casing spacers are attached to the carrier pipe prior to insertion into the casing to provide encasement and grade adjustments. The carrier pipe can be pushed in by hand or with a boring/jacking machine or by pulling the pipe with a winch.

The carrier pipe and joints are designed to withstand the full range of internal operating pressures and external pressures from overburden and live loads. The internal operating pressures are the maximum fluid and/or gas pressures during the operational phase of the pipe. The external pressures are the loads from groundwater, native soil and bedrock, live loads, stray electrical currents, and hydrocarbon contamination. The need for internal and external coatings and linings, as well as active corrosion protection systems, is a function of the pipe material, the conveyed liquid, and the surrounding environment. If carrier pipes are installed within a permanent steel casing, the designer may consider requiring that the annulus between the pipes be filled to reduce pipe movement, ground movements, and/or pipe corrosion.

If the product pipe is directly jacked into place and a temporary or permanent casing is not used, the pipe should be designed to withstand all long-term external soil, groundwater, and live loads. Most state and federal transportation agencies and railway owners have established specifications detailing pipe design criteria, including pipe material requirements, for trenchless crossings, which should be followed during design. In some cases, the product pipe may need to be installed within a permanent concrete encasement or steel casing to satisfy these specifications.

5.2.3 Structural Design of Jacking/Carrier Pipe

From a structural point of view, the pipe has to withstand design/service live (long-term) loads and construction loads. Identifying and quantifying the loads that the pipe should withstand are the first steps in designing the pipe. Construction loads are discussed in Chapter 6. The design loads are broken down into two components: live load and dead load, as shown in Fig. 5-1. The worst loading conditions for these loads are when the pipe is empty. The live loads are loads at the ground surface from trucks and cars on a highway, trains on a railroad, aircraft on an airfield, and so on. The dead loads are permanent loads that do not change as a function of time; they are the overburden pressure of the soil above the pipe and the pipe's own weight. For pressure pipes, the stresses from pressurized liquid or gas must be considered.

5.2.4 Soil Dead Load/Overburden Pressure

The pipe must be stiff to resist the design loads within a design deflection limit; the soil loads depend on soil properties, groundwater elevation, and the depth of the installation. The calculations typically

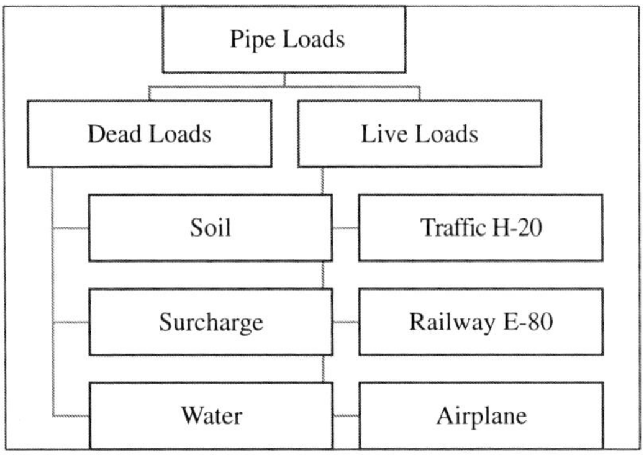

Fig. 5-1. Loads on a pipe
Source: Atalah (1996)

assume an infinitely long pipe in a uniform material; thus, loadings are usually determined as a load per linear foot. A simple assessment of vertical soil load is to use the weight of the soil prism based on the width of the outside diameter of the pipe and the depth of cover times the unit weight of the soil. This load is realistic for shallower installations, but it is a conservative loading for deeper installations because it does not take into consideration soil arching in granular soil or soil strength in cohesive soil. In shallower pipes, the earth loads per unit of length on the casing (P) may be estimated using Eq. (5-1):

$$P = \gamma * H * D \tag{5-1}$$

where γ = soil total unit weight density; H = depth of cover; and D = outside diameter of the pipe.

Since HAB is not recommended below the GWT, the buoyancy effect is not considered. For deeper casings, the modified Marston's formula, used to calculate the soil loads on a pipe installed using the open-cut technique, can be used.

In cohesionless soils, Terzaghi's silo theory, which considers the soil arching action, can be used to estimate the normal pressure acting on the pipe from the vertical load. Eq. (5-2) gives the normal pressure on the pipe (P) as a function of soil density (w), the depth of cover to the top of the pipe (H), the vertical live load at the pipe level (P_L), and the coefficient of soil load (k), which is calculated by Eq. (5-3), which is a soil arching factor based on the mechanical properties of the granular soil, the depth of cover, and the diameter of the pipe.

$$P = k(w * H + P_L) \tag{5-2}$$

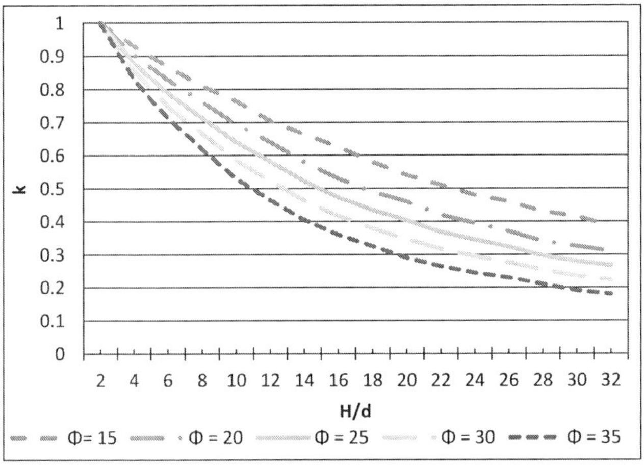

Fig. 5-2. Reduction factor according to ATV A 161
Source: Atalah (1996)

$$k = \frac{1 - e^{-2K \times \tan\delta \times H/B}}{2K \times \tan\delta \times H/B} \tag{5-3}$$

where $K = \tan^2(45 - \phi/2)$, soil's lateral pressure coefficient; δ = angle of wall friction between pipe and soil; B = influence width above the pipe; and ϕ = angle of internal friction in degrees. There are multiple approaches to determine B, all of which have stronger and weaker arguments for usage. The original derivation for circular shapes uses the outside diameter of the pipe as equal to B. Additionally, Terzaghi applied arching reduction when the depth of cover above the top of the pipe was at $1.5B$ or deeper (Proctor and White 1977).

In Germany, the values of these variables are $K = 0.5$, $\delta = 0.5\phi$ (angle of internal friction of soil), and $B = 1.73d$ (the outside diameter of the pipe). Fig. 5-2 presents the value of k as a function of ϕ and the ratio H/d for $K = 0.5$. However, in Japan, $K = 1$, $\delta = \phi$, and $B = d[0.5 + \tan(45 - \phi/2)]$ are used in Eq. (5-3). The values of Terzaghi's variables in Germany and Japan are shown in Table 5-1.

Table 5-1. Terzaghi's Variables for K, δ, and B in Germany and Japan

	K	δ	B
Germany	0.5	0.5Φ	$1.73d$
Japan	1	0.5Φ	$d[0.5 + \tan(45 - \Phi)]$

Note: Φ = angle of internal friction; d = outside pipe diameter.
Source: Atalah (1996).

Table 5-2. Live Load Transferred to Pipe

Depth of Cover (ft)	Highway (H-20) (psi)	Railway (E-80) (psi)	Airport (psi)
1	12.5	N/A	N/A
2	5.56	26.39	13.14
3	4.17	23.61	12.28
4	2.78	18.4	11.27
5	1.74	16.67	10.09
6	1.39	15.63	8.79
7	1.22	12.15	7.85
8	Negligible	11.11	6.93
10	Negligible	7.64	6.09
12	Negligible	5.56	4.76
14	Negligible	4.17	3.06
16	Negligible	3.47	2.29
18	Negligible	2.78	1.91
20	Negligible	2.08	1.53
22	Negligible	1.91	1.14
24	Negligible	1.74	1.05
26	Negligible	1.39	Negligible
28	Negligible	1.04	Negligible
30	Negligible	0.69	Negligible

Source: American Lifelines Alliance (2001).

5.2.5 Soil Live Load

Table 5-2 presents the transferred load to the pipe under various types of live highway, railway, and airfield loads under various depths of cover. The table uses an impact factor of 1.5 to account for bumps and irregularities in the travel surface (American Lifelines Alliance 2001).

If the pavement is designed for heavy-duty traffic, the load pressure of the truck wheel is usually reduced so that the transmitted live load to the pipe is negligible. If the pavement is designed for light traffic but subjected to heavy truck traffic, the jacking pipe should be designed to withstand the live highway loads.

Standard vehicle live loads have been established by AASHTO standard specifications for use in the design of bridges and underground structures subjected to heavy traffic. The commonly specified truck loading is the H-20, which is a vehicle with a standard gross weight of 20 tons with two axles spaced at 14 ft (4.3 m) and designated as a highway truck or "H" truck. The weight of the front axle is 20% of the gross vehicle weight, and the weight of the rear axle is 80% of the gross vehicle weight (Jaramilla and Huo 2005).

For the load transmitted to a pipe installed under railroad tracks, the weight of the track structure including the ballast is considered a uniformly distributed load. The American Railway Engineering and Maintenance-of-Way Association (AREMA) typically recommends a Cooper E-80 loading. The loading requirements of AREMA should also be considered in addition to other requirements of the specific railway owner.

Airport loadings are defined by the Federal Aviation Administration (FAA). These loadings are obtained in FAA design manuals.

5.3 PIT OR SHAFT CONSIDERATIONS

A HAB operation requires an entry (jacking) and exit (receiving) pit. The jacking pit for HAB should be large enough to facilitate the boring operation. The pit must be large enough to accommodate the HAB machine, the track assembly, the thrust block, and the hydraulics to provide the jacking force; the pit should be large enough to safely accommodate the working crew. The pit should be properly shored to provide a safe working environment. Design of the pit is the contractor's responsibility (see Chapter 6 for details).

The jacking and receiving pits should be located at a safe distance from existing structures to avoid any hazards to the structures or the public. The excavation support system must be capable of withstanding static and dynamic loads, including loading from construction equipment, traffic, and stockpiled materials. The excavation support system should meet all applicable OSHA and local safety requirements. Each pit should also be designed with a fall protection system around the pit perimeter and a suitable access system that meets OSHA requirements.

5.4 GROUNDWATER CONSIDERATIONS

As discussed in Chapter 2, the HAB operation does not work well below the GWT. Therefore, lowering the GWT to a depth of at least 1 ft below the pit subgrade is recommended. The spacing, depth, and capacity of the dewatering system will depend on the length and depth of the alignment, the soil stratigraphy, and the groundwater conditions. If dewatering is not feasible, groundwater control through ground modification may be warranted. If dewatering is needed, the designer and the geotechnical engineer must provide the contractor with sufficient information to design the dewatering system. The designer should consider the potential of negative impacts on adjacent structures being caused by dewatering.

If the presence of shallow or perched groundwater is detected during pit excavation, a watertight excavation support system or a groundwater removal system, along with support of excavation, is required. The designer should consider the ability to control groundwater at the tool entry point as well. In addition, the surface grade adjacent to the work pits should be sloped to divert surface runoff away from the pits.

5.5 SITE-SPECIFIC DESIGN FOR RISK MANAGEMENT

This section provides typical investigation techniques and guidelines that the design team can use to identify and assess site hazards. The potential impact on safety, cost, schedule, and environment determines the need for and the extent of testing for the presence of these hazards in the site-specific investigation. The results of the investigation are then assessed and used as a basis for design to manage the risks and mitigation measures associated with the hazards.

Typical site investigations may include

- Desktop study,
- Geotechnical investigation,
- Utility surveys (both aerial and subsurface ground truthing by potholing),
- Investigation of traffic flow and access for vehicles and pedestrians,
- Environmental investigation,
- Investigation of flood zones, and
- Seismic investigation.

All site investigation information available to the owner should be detailed in the bidding documents (mainly in the geotechnical or project design report). Information about the type of soil, the groundwater conditions, the water table, and the existing utilities is to be detailed in these documents. Generally accepted practice is to complete a test boring and determine groundwater elevations at each pit location.

5.5.1 Subsurface Utility Engineering

Subsurface data include existing utilities and geotechnical conditions. Subsurface utility investigation can be initiated once the general site conditions and the surface survey data have been obtained. A critical aspect of subsurface investigation is to locate the existing utilities within 50 ft of the proposed installation. The located utilities along the proposed project bore paths should be plotted (plan and profile) to avoid potential underground utility conflicts or obstruction to the HAB operation. This process is crucial with roadways and railroad crossings, for which the potential and costs of hitting a utility are high. The impact costs may include unplanned, high-cost repairs; disruption to traffic and commerce; personnel injury; and negative public relations. The designer traditionally requires the contractor to confirm the location of adjacent utilities prior to construction through nondestructive (e.g., soft-dig) methods.

The Federal Highway Administration (FHWA) provides general guidelines for the subsurface utility engineering (SUE) processes and procedures needed to accurately identify subsurface utility data for highway plans and permits during the development of a highway crossing project.

The advantages of SUE include

- Developing advance planning procedures and contingency planning,
- Avoiding conflicts with existing utilities during boring operations,
- Reducing schedule delays by eliminating unforeseen conflicts with the existing utilities,
- Minimizing the potential for cost overrun attributable to unexpected utility conflict,
- Minimizing the potential for service interruptions and inconvenience to the public, and
- Increasing the potential for project success.

5.5.1.1 Utility Surveys All utility information (aerial and subsurface) needs to be collected in connection with the project alignment. The information should be collected in accordance with ASCE/CI 38-02, *Standard Guidelines for the Collection and Depiction of Existing Subsurface Utility Data* (ASCE 2002). During the design phase, the designer is responsible for selecting the bore alignment and pit locations that minimize conflict with, and risk of damage to, existing utilities. At this stage of the project, knowing the existence and exact location of any utility within the construction zone is necessary for success.

Utilities should be shown on the contract drawings. Information such as that found in ASCE/CI 38-02 (achieved utility quality level, diameter, depth, material type, owner, trenched cross section, and backfill material used) should be included, if available. If the utility expectancy is complex and/or the risk associated with striking an unforeseen utility is high, using the procedures set forth in ASCE/CI 38-02 to obtain QL-B and QL-A utility data is prudent. The use of a "design" ticket from a state's one-call service can only result in QL-C or QL-D information.

Utilities that are located in proximity to the bore path may require special protection, rerouting, and/or support during construction. If utilities conflict with the new pipeline or shafts, relocation plans should be prepared. Utilities can be relocated in advance of or as part of the project. The remaining utilities in close proximity to the alignment require protection and monitoring during construction. The evaluation of settlement and heave risks may be appropriate, depending on clearances and ground conditions.

5.5.2 Geotechnical Investigation

The design engineer should endeavor to provide the HAB contractor with detailed geotechnical information that includes soil and ground descriptions, soil properties, and anticipated ground behavior, as this information relates to the proposed excavation method. The contractor's ability to match the appropriate HAB system components (cutterhead configuration, HAB machine, augers, etc.) to the ground conditions is heavily influenced by the quality of the information presented in the geotechnical soils report.

Detailed geotechnical investigations should be used to obtain factual information about the distribution and engineering characteristics of soil, rock, and groundwater at the site to the extent required to produce a sound, safe, and economical design and allow reliable construction planning for the work. See Chapter 4 for details of the investigative methods and subsurface conditions that are considered appropriate for auger boring.

5.5.3 Traffic Flow and Pedestrian Access

As indicated earlier in the planning phase, traffic information is critical to the determination of staging areas and shaft locations (see Section 3.2.3). The nature, complexity, and setting of the proposed construction greatly determine the extent and nature of the data required by the designer. Strategic roadway occupancy locations are an important part of the design and construction phase of any project. It is prudent for the design engineer to refer to the FHWA's *Manual on Uniform Traffic Control Devices* (FHWA 2009) for traffic control on public streets and private roads.

The most undesirable alignment location is at the intersection of two roadways. This location interferes with vehicles turning into, out of, and through the intersection and therefore causes traffic delays. Roadway occupancy within 50 ft (15 m) of an intersection is also undesirable. The second most undesirable location is where the boring activities completely occupy the width of the roadway, thus closing all vehicle traffic from the roadway.

The most desirable location for roadway occupancy would be approximately midblock, between two intersections, and to one side of the roadway. This location would allow vehicle traffic to occupy most of the roadway, with little or no disruption. The second most desirable location for roadway occupancy would be an area at least 100 ft (30 m) from an intersection and occupying the middle of the roadway. This location does not generally interfere with turning movements from and through the nearby intersection. In addition, vehicle through traffic on the roadway can pass by the construction area with minimal interference with traffic flow.

In addition to minimizing disruption to vehicle traffic, disruption to commercial/business activities must also be considered and evaluated. Locations of abutters' loading docks, as well as the locations of entrances to parking areas and points of public access, must be known when the selection of a roadway occupancy area is being considered. An abutter's loading dock is critical to business operations and activities. To disrupt the ability of an abutter to conduct business would cause economic loss to that business. Disruption of access to an abutter's parking area is also undesirable, although not as important as disruption to the abutter's loading docks. To the extent possible, the inconvenience costs to the abutter in terms of delays and disruption to its business activities and employees should be minimized. When employees are required to seek alternate parking areas, their personal lives and work-related commuting activities are affected.

5.5.4 Sensitive Environments

Sensitive environments include wetlands, rivers, streams, endangered species habitat, and areas of designated historic or archaeological significance. It is the designer's responsibility to identify the limits of permitted disturbance for sensitive environments in the contract documents and incorporate measures to protect these areas. The contract documents should state the acceptable corrective actions required to mitigate the negative impact if the permitted conditions are violated.

5.5.5 Contaminated Media

Contaminated media are those materials that may present risk to the environment or worker health and safety and that may require special handling and disposal procedures. Contaminated media may include chemical contamination (soil and groundwater) and asbestos related to past site use, as well as naturally occurring contaminants, such as organic or gassy formations and hydrocarbons.

If evidence of contaminated media is revealed during site background research or subsurface investigations, this information should be made available to the contractor. The contractor should also be required to prepare a site-specific health and safety plan and a materials handling and disposal plan to ensure that state and federal regulations and shipping requirements are met. In gassy ground, the contractor should be required to prepare a work space ventilation and monitoring plan.

For contaminated ground, locating the pits outside the contaminated area and then boring through the contaminated area reduces the volume of contaminated spoils to the excavated bore volume. When excavating impacted ground, the volume of contaminated ground used in the design should include that of the pits and the bore.

5.5.6 Obstructions: Rocks, Boulders, Wood, and Artificial Debris

Major obstructions for an auger boring operation can be nested cobbles, boulders, rocks, wood, and debris occurring within the bore path. A preliminary investigation should be conducted to determine the location of any such obstruction to the extent possible. If the potential for obstruction is rated high, alternate routes or bore paths should be considered to avoid project failure. If alternate bore paths are not available, the diameter of the casing may be increased to enable the removal of the anticipated obstructions through the augers, permit the retraction of the augers, or provide personnel entry into the casing for intervention. Generally, an obstruction less than one-third of the diameter of the casing is the largest obstruction that can be removed through the augers. Costs associated with the removal of these obstructions should be considered extra work, depending on the scope of work. Other methods, such as hand mining, should also be considered on the basis of the anticipated soil conditions.

5.5.7 Differing Site Conditions

If site conditions are significantly different from those described in the contract documents and the contractor can demonstrate that the differing site conditions (DSCs), adversely affected the work, the contractor may be entitled to compensation for the delay, impact, and level of effort required to overcome these DSCs. The design engineer should include the DSC clause in the contract documents. A risk-sharing strategy, such as one recommended by the *Geotechnical Baseline Reports for Construction* (USE 2007) is strongly recommended. Some examples of DSCs include the unanticipated occurrence of one or more of the following:

- Boulders,
- Natural or artificial obstructions,
- Bedrock (within a planned soil bore),
- Soil (within a planned rock bore),
- Stable soil conditions that change to unstable flowing sand conditions,
- Cemented soils,
- Shallow or perched groundwater,
- Subsurface contamination,
- Conflicting utilities, and
- Archaeologic discovery.

Two types of changed condition claims are generally accepted within the trenchless industry:

- Type 1: The conditions encountered at the site differ materially from those indicated in the contract documents.
- Type 2: The conditions encountered at the site differ materially from those normally encountered in the general site vicinity.

5.5.8 Flood Zones

Floods, storm surges, and flash flooding conditions should be considered when identifying and selecting pit locations and bore alignments during design. Designing for potential high water conditions can minimize flooding of the pits and damage to the auger boring equipment, prevent pollution from petroleum products, and protect personnel. For HAB projects in flood-prone areas, the contract document should indicate both the flooding season and the window outside of the flooding season during which construction of the project must be completed.

5.5.9 Seismic Conditions

Seismic conditions, including fault offset and related ground responses, should be considered in pit location selection, bore alignment, jacking and receiving pit design, and utility service installation. The seismic considerations include the protection of personnel and the minimization of damage from fault offset, ground shaking, landslides, liquefaction, and tsunamis. The assessment of seismic risk could include the review of information from the USGS, applicable state geological surveys, and local knowledge to determine the location, style, extent, recurrence interval, and probability of seismic conditions relative to the project.

5.6 CONTRACT DOCUMENTS

After completion of the design, the bid/contract documents are prepared by the owner and engineer. The contract documents should provide all the necessary information for the contractor to prepare bids, plan the work, and complete the installation. It is important that individuals with experience in the specified methods and technology prepare and review the contract documents, which must be complete, clear, and concise. These documents may include

- Scope of work,
- General and special conditions,
- Drawings,
- Technical specifications,
- Geotechnical information, and
- Bond forms and requirements.

In addition to the general conditions of the adopted contact documents, the scope of work and the special conditions of the job must be clearly stated. The contract documents should provide general information regarding the type of job, the amount of work, and other relevant data for the contractor, in addition to special information about specific tasks or characteristics of the job. Any other information available about the jobsite should be included in the contract documents.

Contract documents must include all relevant drawings for the site. Care should be taken in detailing special conditions and tasks on the job. The project drawings should provide information about the jobsite conditions; the available work space; the limits of disturbance, erosion, and sediment controls; the depth of installation; the length of the pipe to be bored; and the orientation of the boring. The drawings should also show all the known utilities near the proposed pipeline.

Detailed specifications for the auger boring installation should be included in the contract documents. The type of specification may depend on the preference of the engineer or the owner. Technical details, such as the diameter of the pipe, the pipe material, the method of installing the pipe, and the soil conditions should be included in the documents. Any other technical requirements relevant to the job should be made available to the contractor. At a minimum, the technical specification for auger boring should include

- Minimum machine requirements;
- Required accuracy in line and grade of the bore;
- Invert elevations, casing diameter, pipe wall thickness, length, grade, and strength;
- Type of joining system;
- Product pipe details;
- Blocking or spacers for carrier (product pipe) installations;
- Grouting requirements; and
- Site restoration requirements.

All geological and geotechnical information available to the owner should be detailed in the geotechnical report to include information on the type of soil, groundwater conditions, and water table (see Section 4.5).

5.7 CONTRACTOR PREQUALIFICATION

Auger boring installations require contractors with specialized equipment, expertise, and personnel. It is recommended that the engineer and owner establish minimum experience requirements for the auger boring contractor and key personnel in the contract documents. Specifically, the bidding contractor should be required to demonstrate experience with auger boring in similar ground conditions and using similar pipe materials as the proposed installation.

Contracting companies considered in the prequalification phase should be required to provide information on the type of organization, the type of works undertaken, and the qualifications of personnel involved in the project during the first part of the program. The general information requested may include

- Name and type of organization (e.g., corporate, partnership, etc.);
- Type of contracting business;
- Qualification of key individuals working for the company;
- List of completed projects;
- List of affiliate, partner, or subsidiary companies;
- Organizational setup within the company; and
- Completion rate of the awarded projects.

Contractors should be required to provide information on their equipment and the capacity and age of that equipment. The experience of the firm and its key personnel should be considered in the prequalification requirements with respect to the following:

- List of equipment owned, along with the name, description, quantity, and capacity of each piece;
- Types of work performed;
- Type of work performed using own crew; and
- Safety records.

REFERENCES

American Lifelines Alliance. (2001). "Guideline for the design of buried steel pipe." ASCE, Reston, VA.

ASCE. (2002). "Standard guidelines for the collection and depiction of existing subsurface utility data." *ASCE/CI 38-02*, New York.

Atalah, A. (1996). "Design of microtunneling and jacking pipe." *Pipeline Crossings 1996*, L. F. Catalano, ed., ASCE, New York, 395–402.

FHWA. (2009). "Manual on uniform traffic control devices for streets and highways." Washington, DC.

Jaramilla, B., and Huo, S. (2005). "Looking to load and resistance factor rating." <http://www.fhwa.dot.gov/publications/publicroads/05jul/09.cfm> (Oct. 9, 2015).

Proctor, R. V., and White, T. L. (1977). "Earth tunneling with steel supports." Commercial Shearing, Inc., Youngstown, OH.

USE. (2007). "Geotechnical baseline reports for construction: Suggested guidelines." ASCE, Reston, VA.

CHAPTER 6
CONSTRUCTION CONSIDERATIONS

The construction considerations chapter covers the activities that are usually conducted by the HAB contractor. During the bidding phase, the contractor is expected to review/evaluate the contract document, the geotechnical reports, and the soil conditions to develop behavior expectations for the surface and subsurface environments during the boring phase. Accordingly, the contractor determines the appropriate auger boring system, cutterhead, tooling, and other ancillary support systems required on the basis of the contractor's understanding of the anticipated subsurface condition. After the owner accepts the contractor's bid and awards the contract, the contractor embarks on the activities described in this chapter. Each auger boring project has certain risks and complexities associated with it; therefore, the judgment of the construction team is crucial in determining the level of preconstruction activities/details. This chapter is structured as follows: preconstruction activities, construction activities, and potential solutions to what-if scenarios.

6.1 PRECONSTRUCTION ACTIVITIES

Once the owner issues the notice to proceed, the contractor collects more information about the surface and subsurface conditions of the site and validates/revises the contractor's assumptions against the new collected information. Then, the contractor starts to prepare the needed submittals for the project and obtain the essential permits, easements, and rights-of-way from the applicable permitting authorities. The contractor proceeds further with the following preconstruction activities:

- Investigating and confirming the location of underground utilities;
- Planning the jobsite layout;

- Procuring the needed material, equipment, and accessories; and
- Developing the risk management plan.

6.1.1 Permitting

For work on the project to run smoothly, permits and rights-of-way should be procured well in advance. The working site should be limited to the easements and/or rights-of-way, and if additional space is needed into adjacent properties, permission should be obtained from the property owners. The boring jobs often require the contractor to obtain permits that allow it to execute the work. Below are some of the permits that may be required:

- Right-of-way. The right-of-way permit should be obtained from the pipe owner and/or the owner of the roadway, railway, runway, and so on, as needed.
- Traffic control.
- Water disposal. When dewatering is needed, there could be permitting requirements for the disposal of the water (these requirements could be from EPA, the state, and/or local authorities).
- Any other permits stated in the scope of work in the contract documents.

6.1.2 Submittals and Procurement

The timely procurement of long lead items is critical to the timely completion of the project; therefore, planning of the submittal preparation, review, and approval activities is needed. Only after the owner has approved the submittals of these items can the contractor place the order for these items. The coordination of the submittal, procurement, and construction processes is paramount for timely completion. Some of the items with long lead times are the casing, carrier pipes, and spacers. As the augers and the cutterhead are critical parts of the means and method of construction, the owner's approval of the types selected is advisable.

The contractor is required to prepare and provide detailed submittals of the auger boring project to ensure that the contract and design requirements are met. The extent of the needed submittals depends on the specific risk(s) and complexities associated with the project. Some examples of additional required submittals are

- Site layout plan;
- Casing pipe manufacturer brochures and certification tests;
- Brochures, specifications, and operating manual of the HAB machine;
- Brochures, specifications, and operating manual of the lubricating system;
- Steering system to be used and the line and grade control plan;
- Qualifications of the workers who will be involved in the project;

- Pit design, shoring systems, and other construction considerations for the pits;
- Dewatering plan;
- Spoils management plan;
- Contingency plan;
- Work safety plan;
- Daily progress reports, including machine performance; and
- Risk management plan.

6.1.3 HAB Work Plan

The work plan should include the following items:

- Permits,
- Excavation techniques,
- Design responsibilities,
- Supervision,
- Inspection,
- Protection of adjacent structures,
- Public safety,
- Ventilation requirements,
- Spoils removal from pit (skid steer, backhoe, crane and bucket, vacuum, etc.), and
- Excavated material management and disposal.

A contingency provision for the removal of possibly contaminated soil can be included in the plan. In addition, contingency plans should be provided for unexpected subsurface conditions and the occurrence of obstructions. For pit excavation and boring operations, all OSHA rules, and those in the manufacturer's equipment safety and operations manual, should be followed.

In general, all DOT and railroad guidelines for the depth of cover should be followed during both planning and installation.

6.1.4 Dewatering Considerations

Auger boring should not be used in silty and sandy soils below the GWT. However, if auger boring must be used below the GWT in sandy or silty soils, an elaborate dewatering system, such as a well point system, deep wells, or a larger sump and pump system, is recommended. As is the case with every dewatering system, the contractor should take all necessary measures (such as gravel and sand bedding or wrapping filter fabric around the well or the sump barrel) to prevent the migration of the soil particles. If dewatering needs arise, safe and proper flow discharge plans are required; therefore, a suitable discharge in compliance with the EPA requirements is needed. If a sump pump is used, the preliminary treatment of flow to reduce the sediments may be needed before discharging into the water streams.

The receiving and jacking shafts should be dry during installation to avoid disturbing the subgrade in the shaft. Therefore, if rain is expected or the pipe invert is slightly below the GWT in clay soil, the installation of a dewatering sump pump at one corner of the shaft is needed. Ditches crossing the shaft, sloped toward the sump, lined with filter fabric, and filled with gravel may be needed to direct the water toward the sump.

6.2 JOBSITE LAYOUT PLAN

The jobsite layout plan reflects the intended method of construction and addresses the above-mentioned considerations. The contractor does not start boring before the engineer reviews and approves the jobsite layout plan. If the site conditions mandate a deviation from the approved plan and submittal, the contractor should notify the project owner and obtain the approval of the owner prior to the deviation from the plan.

The work space should provide enough room for the safe operation of equipment and for the storage of the needed materials safely and economically. The work space includes both the space available on the site and the space inside the pit. The space inside the pit should allow the free and safe movement of personnel and equipment. The pit should be large enough to permit the personnel to work around the pipes to set, attach, weld, and detach casing pipes. The space should also be large enough for the crane/excavator to lower the casing into the pit and remove the spoils from the boring. Typically, casings are 10, 20, or 40 ft (3, 6, or 12 m) long. The entry shaft is commonly 30–37 ft (9–11 m) long × 8–14 ft (2.5–4 m) wide if the casings are 20 ft (6 m) long. The surface working area is usually about 75 × 150 ft (23 × 46 m) with a minimum area of 30 × 82 ft (9 × 25 m) (Najafi 2013).

The jobsite layout should be well planned to facilitate the storage and use of various required materials and equipment safely and efficiently with minimum handling and rehandling. The layout should be prepared and approved by the engineer well in advance of the commencement of work on the site. The layout should provide adequate space for the required equipment, material storage, and job trailers around the jacking shaft. The materials and equipment required for the job should be located close to the jacking shaft to reduce handling time and efforts. The material items include augers, casings, pipes, excavated material, and carrier pipes. The equipment may include lubrication mixing and pumping system, a crane/excavator, the HAB machine, and the welding machine. The crane/excavator is positioned with enough room to maneuver around the shaft and lift its loads safely without tipping or damaging the shoring systems or the slope for the jacking shaft.

Fig. 6-1. Typical jobsite

Fig. 6-1 shows a typical jobsite layout, which allows for

- Pit locations;
- Storage of materials, such as pipe sections, augers, and bentonite bags;
- Lubrication system; and
- Support equipment, such as backhoe, forklift, and crane.

6.3 UNDERGROUND UTILITIES INVESTIGATION

The contractor should do its due diligence to identify, locate, and verify nearby underground utilities prior to digging the shafts and the boring drive. Prior to the boring operations, the contractor should call the local one-call locating center to have representatives of the nearby utilities come to the site and detect underground utilities by nondestructive means. This process involves the use of several tools and should include a utility locator and utility maps. After marking the existing utilities on the ground surface, careful excavation is used to verify the depth, location, direction, and diameter of the utility. Manual excavation may be needed for the last few inches from the existing utilities to avoid damaging those utilities.

Utilities in moderate structural condition that are further than two times the diameter of the casing from the boring head are unlikely to be damaged but should be monitored. Utilities that are closer to the boring head than this distance or are in poor structural condition should be exposed, protected, supported, and/or rerouted as needed prior to the boring operation. It is considered good practice to expose the utility in the bore path before the bore and leave the utility exposed until the bore passes the utility. Care should be taken to avoid damage, as some of the utilities may be unprotected or the utility may not be in the expected location.

Vacuum excavation (potholing) is an excellent tool to expose utilities with minimum surface excavation and minimum risk to the existing utility. The technique can be used to create 12- to 18-in.-diameter (300–450 mm) holes to confirm, physically, the position and depth of an underground utility. If a utility is under a pavement, a rotary core drill is used to cut a hole in the road pavement, and then compressed air jets and/or high-pressure water jets are used to advance the excavation. This excavation process does not normally damage an existing utility, and the hole size in the street pavement is kept to a minimum and thus can easily be repaired. This procedure can only be used to confirm the position of known utilities or previously located utilities.

6.4 CONSTRUCTION MEANS AND METHODS

In light of the additional acquired information (surface and subsurface), it is helpful to analyze thoroughly the suitability of the HAB method for the proposed site conditions. A working understanding of the method is an important aspect of the construction process.

The most critical part of any auger boring operation is matching the tools and system components to the expected soil conditions. Some of these system tools are the cutterhead, the augers, the casing, the jacking shaft, the thrust block, the lubrication fluids, the grade monitoring system, and the steering head. HAB can be used in a wide range of soil, from dry sand to clay to solid rock formation. However, each subsurface condition requires specific tooling. Each specific tooling has a limited range of subsurface conditions in which the tool can function effectively and efficiently. Encountering conditions outside of that planned range of tooling capability can negatively impact the construction method or, more often, can require a complete retooling, along with the associated schedule and cost impacts. Some specific conditions that are known to have significant tooling impacts are potential obstructions, such as boulders, artificial obstructions, and mixed-face conditions. These conditions can often be handled successfully, but they may pose a challenge to HAB projects.

6.4.1 Cutterheads

Cutterheads are available in the marketplace in a variety of configurations and steel alloys. It is crucial that the cutterhead be suitable for the expected soil type. Refer to Chapter 2 for more information on cutterheads.

6.4.2 Auger Options

It is recommended that all bores be made with a complete string of full-sized auger sections. However, under conditions in which the auger loading is light and the spoils move easily in the casing, the lead sections of a full-sized auger, followed by smaller sections, can be used. However, a smaller-diameter auger should never be used in the lead section of the casing. The step-down in auger size should not exceed one-third of the casing size to avoid decreases in efficiency, as not all the spoils are removed from the casing when the smaller-diameter auger is being used at the rate of excavation. The use of a smaller-diameter auger results in the auger rotating through more revolutions to remove the volume of soil being excavated at the face, which again results in the rotation of the auger without forward advancement. Other factors in the use of augers smaller than the casing are bending and torque. An undersized auger creates bending, which results in stresses in the auger stem. A smaller auger will have more windup from the same torque loading than will the full-sized auger. Torque windup pulls the cutterhead back toward the casing and could cause the wing cutters to make contact with the casing, further increasing torque and causing even more damage.

6.5 JACKING AND RECEIVING PITS

A HAB operation requires two work shafts: one for boring, called the "jacking shaft," and one for receiving, called the "receiving shaft." The boring pits should be located at a safe distance from existing structures to avoid any hazard to these structures or the public. The distance of the pit from the roadway should be adequate to allow the sloping of the pit if necessary. If sufficient sloping cannot be accomplished, then an earth support system of pit walls should be considered. Enough room for spoils removal and for safe loading and unloading should be provided. Accidents are less likely to occur at sites that are open and kept clear of debris.

Depending on the job conditions, the collected data, traffic conditions, and available easement, the contractor determines the locations of the jacking and receiving shafts. It is preferable to bore from the downstream end to take advantage of gravity in moving excavated material and, more importantly, for water to drain from the inside of the casing if personnel entry is required. Sufficient operations space around the jacking shaft is needed for the following items:

- Storage of materials (such as pipe sections, augers, and bentonite bags),
- Lubrication mixing and pumping system,
- Support equipment (such as backhoe, forklift, and crane),
- Dirt hauling truck or space for piling the excavated material, and
- Additional space for typical construction operations.

The jacking shaft should be large enough to accommodate the auger machine, the rails, a thrust block, and the hydraulics to provide the jacking force. The jacking shaft should be longer than the length of the casing pipe joint, plus the HAB machine, the distribution ring, and the thrust block. The jacking shaft should also be wide enough to allow easy removal of the spoils and provide adequate work space for the workers who will be in the pit (including the welders).

After the bore line is surveyed and stacked at the ground surface, the jacking pit is outlined. To facilitate spoils removal, the pit should be offset from the bore line toward the side where spoils exit the auger. Excavation can begin once the utilities are located and marked. The utilities that are encountered in the pit must be properly supported. An adequate dewatering system must be installed if groundwater is expected.

The bottom of the boring pit should be filled with crushed stone or gravel to create a firm subgrade to support the tracks, the boring machine, the casing, and the auger. Usually, wood planks measuring 2 × 8 in. (50 × 200 mm) × 16 ft (5 m) are placed parallel to the track rail, under the track, for support. A concrete floor may be placed if the bore is of considerable length, size, and duration, and/or soil conditions warrant such placement. A concrete floor is recommended for boring in rock conditions. In all cases, the track support must be set to the proposed grade of the bore. Fig. 6-2 shows two equipment setups.

As the boring machine exerts jacking pressure on the steel casing, the pressure will be transferred to the soil in two locations: (1) along the full outside surface area of the casing through skin friction, pushing the casing forward, and (2) to the soil at the back end of the jacking shaft or the soil underneath the jacking shaft. In the jacking shafts, the jacking force is transferred to the soil via a thrust block (Fig. 6-3), steel shanks driven into

Fig. 6-2. Machine setups in the pit

Fig. 6-3. Simple thrust block

the soil, blanks anchored into the bottom of the jacking shaft, and so on. Regardless of the transfer system, the soil must be able to withstand the stresses coming from the HAB machine. The passive earth pressure of the soil has to exceed the stresses generated by the jacking force with an acceptable factor of safety.

For short runs with smaller-diameter casings, the friction between the track and the soil can provide sufficient counterpressure to the exerted jacking forces. For longer runs with larger-diameter casings, the boring machine and the tracks exert thrust to the back side of the boring pit. A thrust block is installed at the back side of the jacking shaft to distribute the jacking forces uniformly to the supporting soil and counterbalance the thrust exerted by the boring machine. The block can be steel sheeting, steel plate, timber, or concrete. Adequate care should be taken to ensure that the thrust pressure developed by the operation does not affect existing utilities near the bore pit.

Sometimes, instead of the jacking force being transferred to the soil through the thrust block, the force is transferred to the soil through the steel shanks that anchor the track in the soil underneath it. The shearing strength of the shanks should exceed the expected jacking force with a sufficient factor of safety.

In most cases, an exit pit is required at the end of the bore. The safety requirements for an exit pit are the same as for an entrance pit. Unless necessary, no personnel should be allowed into the exit pit during the boring operation. As the casing pipe approaches the exit pit, care should be taken to prevent the collapse of the pit walls.

The possibility of flooding always exists during the boring operation. The location of a pit sump for pumping should be considered during the design of the pit. The sump is placed on the right or left rear, depending on the slope of the pit floor. This placement will move water away from the ejected spoils on the left side of the machine. Additionally, jacking and receiving pits can be used to verify geotechnical information.

6.6 SHORING JACKING AND RECEIVING SHAFTS

It is the responsibility of the contractor to ensure that the pits are constructed according to the OSHA Code of Federal Regulations, Construction Standards for Excavations, 29 CFR, Part 1926, Subpart P. There are specific requirements for pit construction, protection, barricades, traffic control, and installation, as well as for the types of ladders used in the pit and for personal safety equipment. Information can be obtained from the regional Department of Labor office.

The shaft should be properly shored/sloped to prevent it from collapsing. Proper shoring of the jacking and receiving shafts is essential for the safety of the workers and the safety of the nearby utilities and structures. Some of the available means of shoring these shafts are trench boxes (Fig. 6-4), solder piles and lagging, steel sheet piles, and corrugated pipes. The judgment and supervision of a competent person (as defined by OSHA) or a qualified geotechnical engineer are needed to ensure safe shoring.

The shoring must withstand the active horizontal earth pressure resulting from the weight of the soil and the weight of nearby equipment, trucks, rail tracks, and so on. A typical procedure would be

- Using the vertical (soil and live) loads and the soil properties to calculate and draw the earth pressure diagram according to established lateral earth pressure theories in soil mechanics,

Fig. 6-4. Trench box shoring systems

Fig. 6-5. Sloped sides of the jacking shaft
Source: Midwest Mole (2014)

- Calculating the internal forces (bending moment diagram) and stresses that would be sustained by the shoring system, and
- Designing the cross section of the shoring system to withstand the above-mentioned stresses.

If space is available, the walls of the shaft can be sloped instead of providing structural support. Many of the auger boring operations involve crossing railroads and highways, where space is plentiful, the pipe is not too deep, and the alignment is above the GWT. In such cases, the more economical alternative would be to slope the sides of the jacking and receiving shafts, as shown in Fig. 6-5. To avoid slope failure, the resisting forces must exceed the driving forces. Many times, the auger boring contractor's experience leads to the most suitable shoring system.

6.7 CASING PIPE

During the construction phase, the casing must be able to withstand the construction loads, which include the maximum jacking loads, the misalignment of eccentric loads, the construction surcharge loads, and the construction handling loads. These loads can be grouped into the axial jacking force and the transverse loads (normal to the pipe wall). These effective normal loads are the pressure on the outside surface of the pipe from vertical loadings (dead loads and live loads), as discussed in Chapter 5. The axial loads result from the friction resistance along the outside surface of the casing, which is a function of the effective normal loads on the pipe and the friction between the soil and the pipe. The frictional resistance (R) is calculated as follows:

$$R = \pi \times D \times L \times (V \times \mu + C) \qquad (6\text{-}1)$$

where D = outside diameter of the casing; L = length of the casing; V = effective normal stress on the casing); μ = coefficient of friction; and C = soil adhesion. The coefficient of friction between the pipe and the soil depends on the outside surface roughness of the casing, the lubrication, and the soil friction. There are many ways to accurately calculate the dynamic and static coefficient of friction (frequently covered in the literature). However, the simple approach based on friction between steel sheet piles and soils is sufficient (see Tables 6-1 and 6-2).

Soil adhesion is a characteristic of cohesive soil; typical values of soil adhesion are presented in Table 6-2.

There are two techniques to reduce the required jacking force through the pipe and keep the axial stress below the design strength of the pipe:

Table 6-1. Surface Friction Angles and Coefficients

Soil Type	Friction Angle δ	Friction Factor $\mu = \tan\delta$
Sandy gravel, clean	28	0.55
Sandy gravel, silty	23	0.42
Dry medium sand	28	0.55
Damp sand	28	0.55
Saturated sand	26	0.49
Dry silt	28	0.53
Wet silt	20	0.36
Clean sand, silty sand–gravel mixture, single-size hard rock fill	17	0.30
Silty sand, gravel, or sand mixed with silt or clay	14	0.25
Fine sandy silt, nonplastic silt	11	0.20

Source: Atalah et al. (1994), © 1994 NASTT.

Table 6-2. Typical Values for Soil Casing Adhesion

Soil	Soil Cohesion (psi)	Adhesion (psi)
Soft	0–750	0–600
Firm	750–1,500	600–1,500
Stiff	1,500–3,000	0

Source: Atalah et al. (1994), © 1994 NASTT.

(1) oversize cut and (2) lubrication of the outside surface of the pipe. If the soil is highly stable, the wing cutters and reinforcement band overcut the soil during excavation, which reduces the jacking force. Lubrication around the whole perimeter of the pipe and along the whole length of the drive substantially reduces the friction.

6.7.1 Wall Thickness

The interaction between the soil and the pipe has to be considered in the design. As casing pipe is made of steel, it compresses and deforms under the soil pressure. The pipe is checked against all loads at the weakest section of the pipe (the joint).

The maximum axial load that the casing pipe can withstand is calculated using Eq. (6-2):

$$Pmax = \frac{A \times C}{FS} \tag{6-2}$$

where $Pmax$ = maximum axial load the pipe could stand (jacking forces); A = smallest cross-sectional area of the pipe; C = allowable compressive strength of the pipe material; and FS = factor of safety. The jacking forces will be greatest on the sections of pipe closest to the jacking equipment after jacking the whole drive length. These forces are rarely distributed evenly around the pipe's end circumference because the pipe alignment is seldom perfect due to defects in squareness and the mating of joints. Some eccentricity of the axial load will typically occur in the field and result in force concentrations in portions of the pipe ends (joints). Therefore, the use of a reasonable factor of safety to compensate for these imperfections is important. Jacking loads can be reduced by the application of lubricants along the pipe exterior. The impact and effectiveness of lubrication will vary by installation and depend on the subsurface materials present, the annular overcut, the lubricant components, and the injection methods.

Axial loads are often the dominant loads when designing the jacking pipe; however, transverse loads may be critical at shallow depths under heavy live loads or for short lengths of installations. The casing pipe is made of steel and is considered a flexible pipe that follows the flexible pipe design procedures. The casing pipe should be adequately designed to withstand the jacking force from the HAB machine, the overburden pressure, and the potential damage that may be caused by the rotating augers.

The casing wall thickness commonly ranges from 0.25- to 1.5 in. (6 to 38 mm) on the basis of the design loads and the strength of the steel material. The casing should have sufficient strength to withstand the installation loads (friction and the thrust applied by the boring machine). Standard sizes for steel casings are shown in Table 6-3. A casing installed

Table 6-3. Minimum Wall Thickness for Steel Casing Pipe

Nominal Size, in. (mm)	Highway Crossing, in. (mm)	Railroad Crossing, in. (mm)
6–14 (150–355)	0.188 (4.78)	0.25000 (6.35)
16 (465)	0.188 (4.78)	0.28125 (7.14)
18 (450)	0.25 (6.35)	0.31250 (7.94)
20 (510)	0.25 (6.35)	0.34375 (8.73)
22 (560)	0.25 (6.35)	0.34375 (8.73)
24 (600)	0.281 (7.14)	0.37500 (9.53)
26 (660)	0.281 (7.14)	0.40625 (10.32)
28 (710)	0.312 (7.92)	0.43750 (11.11)
30 (750)	0.312 (7.92)	0.46875 (11.91)
32 (815)	0.312 (7.92)	0.50000 (12.70)
34 (865)	0.312 (7.92)	0.53125 (13.49)
36 (900)	0.344 (8.74)	0.53125 (13.49)
38 (965)	0.344 (8.74)	0.56250 (14.29)
40 (1,015)	0.344 (8.74)	0.59375 (15.08)
42 (1,050)	0.344 (8.74)	0.62500 (15.88)
44 (1,120)	0.344 (8.74)	0.65625 (16.67)
46 (1,170)	0.344 (8.74)	0.65625 (16.67)
48 (1,200)	0.344 (8.74)	0.68750 (17.46)
50 (1,270)	0.71875 (18.26)	
52 (1,320)	0.75000 (19.05)	
54 (1,370)	0.78125 (19.84)	
56 (1,420)	0.81250 (20.64)	
58 (1,470)	0.81250 (20.64)	
60 (1,525)	0.84375 (21.43)	
62 (1,575)	0.87500 (22.23)	
64 (1,625)	0.90625 (23.02)	
66 (1,675)	0.93750 (23.81)	
68 (1,725)	0.93750 (23.81)	
70 (1,780)	0.96875 (24.61)	
72 (1,830)	1.00000 (25.40)	

Note: Minimum thicknesses assume at least 4.5 ft (1.37 m) of cover over the top.
Source: Iowa DOT (2009).

below active roadways, runways, and railroads tracks must be designed to withstand equipment and vehicle live loads. Most state and federal transportation agencies, as well as railway owners, have established specifications detailing pipe design criteria for trenchless crossings; these specifications should be consulted during design.

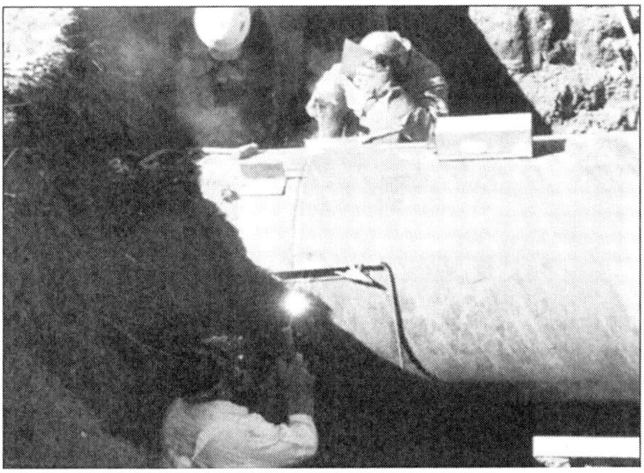

Fig. 6-6. Welding two pipes together

6.7.2 Welding Steel Casing

Welding could be either stick or metal inert gas (MIG) and use either welding sticks or wire. The welding equipment manufacturer's guidelines on maintenance and operation must be followed. The welding must be done by a qualified welder, and all safety standards must be followed. It is important for the operators to be effectively trained in the safe use of the welding equipment before welding the casing. It is advised that all operators refer to ANSI Z49.1, *Safety in Welding, Cutting, and Allied Processes*, which is the ideal standard to guide and ensure safety in welding operations.

For a proper joint that meets the specifications, the pipes must be cut square and beveled to allow for proper welding. Small steel pieces can be welded on one of the casings at two or more positions to ensure that the two pipes form a straight joint. Tacks can then be made at various positions of the joint—say, the 12, 9, 6, and 3 o'clock positions—before proceeding with the welding around the joint, as shown in Fig. 6-6.

Interlocking steel pipes (Permalok), shown in Fig. 6-7, can be used for the casing to provide quick, easy, and permanent joints that eliminate the need for in-field welding and, essentially, increase productivity.

6.7.3 Banding the Casing

The use of a partial band at the head end of the casing is recommended when boring in most soil conditions. The band compacts the soil and creates an overcut that decreases skin friction. The banding process is most effectively utilized in unstable soil conditions when wing cutters are not used. In such instances, the casing is pushed forward without the borehole being overexcavated. Therefore, the soil compacting benefit is more pronounced,

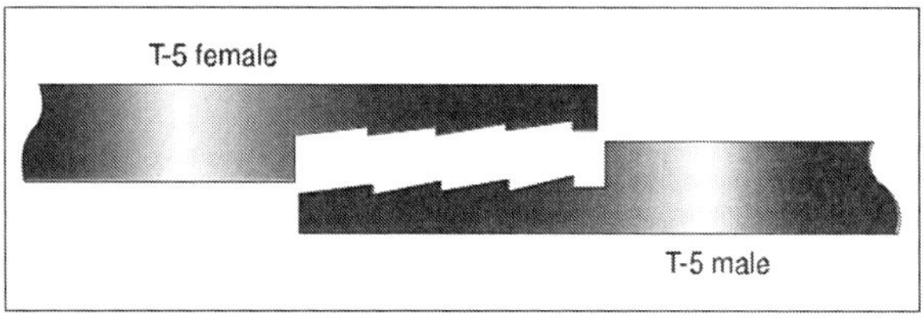

Fig. 6-7. Interlocking pipe joining system
Source: Permalok (2002)

Fig. 6-8. Banding of the casing

as pressure is relieved on the following casing sections. Banding is also beneficial in rock or boulders, as it strengthens the leading edge of the casing.

The placement of the band is a matter of personal experience, but the band is usually 3/8 × 6 in. (10 × 150 mm), rolled to fit the casing. A gap of approximately 8–10 in. (200–250 mm) is left at the bottom. The band is placed so that it leads the casing by about 1/2 in. (13 mm) and is welded securely along the front on the inside and along the rear on the outside. The inside chamfer of the front weld will provide a lifting action for the casing if the thrust is rapid. If wing cutters are used to overcut the casing, a falling action will occur. Fig. 6-8 illustrates a typical banding of the casing.

6.7.4 Installing the Casing

Once the entry/jacking pit has been prepared and the thrust plate/block is installed, the track is lowered into the pit and checked to confirm that the track is on the line and grade of the designed bore. This checking is a critical step in the process, since the system does not allow for significant steering capability; therefore, the importance of setting the pit base and the track on line and grade cannot be overemphasized. Fig. 6-9 shows the setup of the back plate and the tracks in the pit.

Fig. 6-9. Pit setup and setting of machine and tracks

Fig. 6-10. Cutterhead connected to the augers in the lead casing

The machine is then lowered and secured onto the track and positioned at the back of the track. On the ground surface, the HAB crew inserts the auger(s) inside the banded casing and attaches the cutterhead to the lead auger, as shown in Fig. 6-10. The casing is lowered into the pit, and the augers are connected to the front drive. Plates that are welded on the casing and bolts on the master casing pusher are used to secure the casing to the master casing pusher and prevent the casing from rotating during the boring process. The machine is then thrust forward until the cutterhead is in contact with the excavation face.

Collaring is the first operation when beginning a bore. It is the initial entry of the casing or cutterhead into the ground. The objective is to start the entrance of the cutterhead into the ground without lifting the casing out of the saddle. This process involves rotating at low revolutions per minute (RPMs) and using a slow thrust advance. When about 4 ft (1.3 m) of casing has entered the ground, the engine is shut down, the saddle is removed, and the line and grade of the casing are checked. If the casing is not in line with the proposed bore, the casing is removed and the process is repeated. The success

Fig. 6-11. Beginning of a bore

of the bore depends greatly on the line and grade of the first section of the casing. Fig. 6-11 is a picture of the cutterhead at the beginning of a bore.

After the first section of the casing has been successfully installed in the ground, the casing is cleaned by rotating the auger until all the spoils are removed. The machine is shut down, and the auger pin in the spoils chamber is removed. The machine is then restarted, moved to the rear of the track, and again shut down. Then the next section of the casing and auger is lowered into position. The augers at the face are aligned flight to flight, the hexagonal auger joint is coupled, and the auger pin is installed. Steel bar stock, set on edge, is welded on the driven casing at the 11 o'clock and 1 o'clock positions. The casing is then advanced over the auger. The second casing is aligned with the first casing by resting on the welded bars or angles to the installed casing. Before welding, the two pipes must be aligned with straight edges along the top and the sides of the two pipe joints. If the second casing is in line with the first casing and seriously out of line at the machine end, then the first casing is misaligned and must be corrected. Otherwise, an unacceptable bore alignment will result. The first casing should have waterline and lubrication lines already in place. Once the second casing is aligned with the first casing, the two are tacked together and then welded completely. Fig. 6-12 shows two casings being aligned, and Fig. 6-13 shows two casings being welded together. The drive is then coupled to the auger, and the casing is secured to the pusher. Water and lubricant lines, if being used, are added. The machine is then restarted, and the casing is installed. The process is then repeated until the bore is completed.

When the bore is completed, the machine is shut down and the cutterhead is uncoupled from the augers. The machine is then restarted, and the augers are rotated to clean the casing, then pulled back out of the installed

Fig. 6-12. Aligning additional casings

Fig. 6-13. Welding casings together

casing. Fig. 6-14 shows the cutterhead as it emerges at the exit pit, and Fig. 6-15 shows the augers being pulled back out of the installed casing.

6.8 LUBRICATION FLUIDS

The lubrication fluids in HAB applications are used to reduce friction between the casing pipe and the surrounding soils. The lubrication fluids

Fig. 6-14. Cutterhead at the exit pit

Fig. 6-15. Auger pullback

supplier should be consulted about the mixing instructions and concentrations to achieve proper fluidity and characteristics. The selection of lubrication fluids depends on the type of soil through which the pipe will be driven. Generally, in sandy and silty soils, bentonite mixed in water is sufficient to create a wet layer of filter cake (like the consistency of toothpaste) around the casing. Generally, in sticky and swelling clay soils, polymers mixed with other additives are used to prevent the clay from hydrating, swelling, and sticking to the pipe and equipment tooling. Lubricating fluids can significantly reduce the jacking forces (skin friction) on the casing and reduce the risk of ground movement.

Fig. 6-16. Lubrication system setup in the launch pit

The lubrication ingredients are mixed together in a specific sequence, depending on the nature of the mixing water and the lubricants. The fluid is pumped through a manifold at the top of the casing near the boring head and at regular intervals for longer drives. A channel welded to the top of the casing houses the hose that carries the fluid from the lubrication pump to the manifold. The fluid should fall by gravity to lubricate the full circumference of the pipe. Fig. 6-16 shows the setup of a lubricating system in the launch pit.

6.9 CARRIER PIPE INSTALLATION AND GROUTING OF THE ANNULAR SPACE

The carrier (or product) pipe can be of any material, including PVC, HDPE, concrete, and clay. When line and grade accuracy is required, it is important that the carrier pipe be blocked in place to avoid flotation and allow for the grade adjustment of the carrier pipe within the casing, as in Fig. 6-17. Wood blocking, premanufactured spacers, or casing insulators can also be used. The spacers can be plastic, fiberglass, stainless steel, and carbon steel. They can be coated in epoxy, rubber, or other material.

After inserting the blocked carrier in the casing, the grout or sand can be pumped into the annular space between the two pipes. If sand–cement grout is used, care must be taken when grouting small pipes in a large annulus, as the heat of hydration of the sand–cement grouts can damage some plastic pipes. Grout is far denser than most carrier pipes, even if the pipe is capped and filled with water. There is a very real potential of "floating" the carrier pipe if grout is used. Plastic pipe has become larger and lighter in recent years, making it more prone to floating and "hammering." Lightweight cellular grouts are more popular alternatives for this

Fig. 6-17. Wood blocking banded to carrier pipe

application. The cellular grouts have a lower density than sand–cement grouts and therefore reduce the possibility of flotation of the carrier pipe, and they have superior fluidity, allowing for low installation pressure. A compressive strength of 150 psi is usually recommended in most applications (Najafi 2013).

When using sand, it is important to block the pipe down, as sand and pea gravel do not have sufficient strength to hold the carrier pipe down and keep it from floating. This is especially true with larger-diameter pipes. Carrier pipes sometimes displace sand and pea gravel, even if the annular space between the carrier and the casing is entirely full. Care should be taken when blowing sand or pea gravel, as they are abrasive and may damage the pipe. The installation process involves using air to jet the material inside of the casing under high pressure.

6.10 STEERING SYSTEMS (TRACKING, LOCATING, AND GUIDANCE CONSIDERATIONS)

6.10.1 Waterline System (Grade Only)

The waterline system (Dutch level assembly or grade indicator) is the most used device to monitor the grade of the casing pipe as it is being installed. The system uses a water-level sensing head attached to the top of the leading edge of the casing. The level operates in the same way as the sight tube on a boiler. Both ends of the system are vented to ambient pressure. The HAB operator mounts the hose onto the control and indicator board at a convenient point in the jacking shaft. The mounted hose is connected to a water pipe running along the top of the casing, like the one

Fig. 6-18. Steering rods and water pipes held by clips on top of casing

shown in Fig. 6-18. The sensing head is attached to the lead casing just behind the steering head and connected to a pipe that runs on top of the casing.

A water supply line is attached to the grade indicator that feeds the line all the way to the sensing head. The level of water in the indicator board hose shows the level of the valve at the end of the casing as it is pushed into the ground. Fig. 6-19 presents the water pipe setup over the casing connecting to the grade indicator.

A drum, like the one shown in Fig. 6-20, can be used for the water supply (or antifreeze during cold weather). The drum must be kept reasonably full of water throughout the boring operation. The system is then charged, and

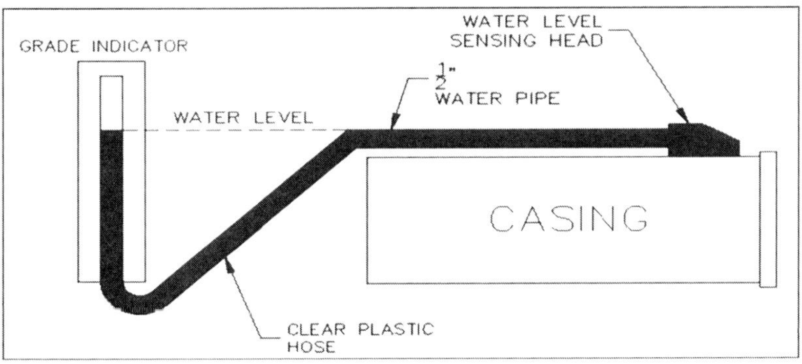

Fig. 6-19. Water pipe setup over the casing

Fig. 6-20. Drum used for water supply (or antifreeze in cold weather) at ground level

the water is allowed to flow through the system until the air is purged from it. At this point, the system is ready for use.

When using this system, the operator should be careful to ensure that the water line is free of air bubbles so that an accurate reading can be made of the elevation. Fig. 6-21 illustrates a sample grade indicator, and more information can be obtained from detailed steering head and water-level manuals provided by equipment manufacturers.

To effectively use a water-level and steering head, the operator should first determine the rise per given length of casing. The grade will be monitored in this interval, following the boring and jacking process.

The level of water on the grade indicator should indicate the level of the sensing head at the end of the casing as the casing is being pushed into the ground. After every predetermined length of jacking, the operation should be halted to check the level (Fig. 6-22). The air is purged from the water system, the level is confirmed, the steering head is adjusted (if required), and finally, the information is recorded on the chart. This process is repeated until the casing is in the ground. The next section is then ready to be installed.

When dealing with successive casings, once the new casing is welded to the installed casing and is attached to the machine, the steering rods and water pipes are connected and the air is purged from the system before the next reading is taken. Fig. 6-22 shows the casing being pushed into the ground with a water hose to check the level connected to it.

6.10.2 Laser- and Other Guided Installations

In the SBU-M and SBU-RH, a laser beam is shot at a target located near the back end of the SBU. Any deviations in line and grade can be noted when the laser beam is off target, and the head can then be steered to the right path. The augers rotate inside a smaller pipe inserted inside the casing

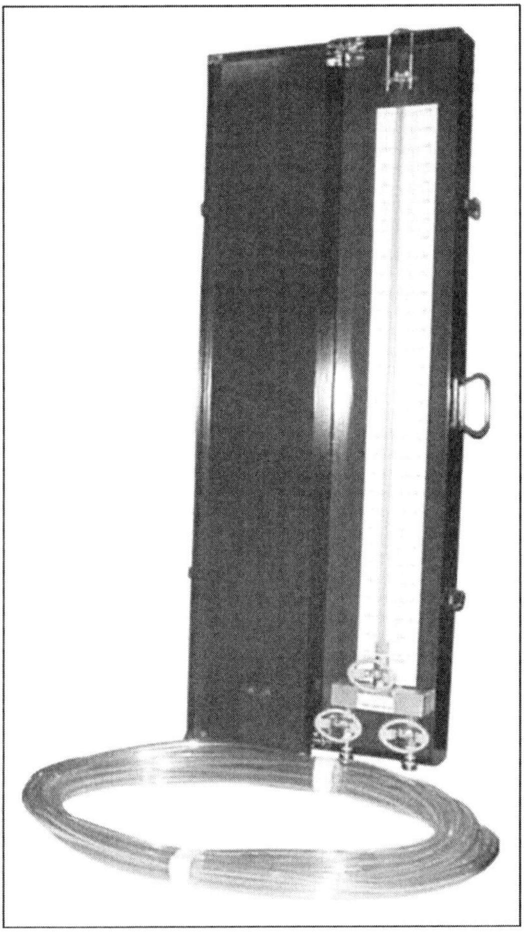

Fig. 6-21. Sample casing grade indicator

pipe. The laser beam runs above the encased augers inside the bigger casing to the target behind the auger head.

6.10.3 Line and Grade Control Systems

Chapter 2 covered various new technologies that can be used to achieve better control of both line and grade (steering the bore if it deviates from the target path). However, the most common method used to steer the bore is a mechanical biaxial hinge incorporated into the lead piece of casing. A set of steering rods (inserted inside a steel pipe welded to the outside of the casing) is used to make line and grade corrections. Miniature gearboxes are used to reduce the effort to turn the rods for line and grade control on long

Fig. 6-22. Casing being pushed into the ground with water hose to check the level connected to it

bores. The leading edge of the casing should be properly prepared, and care should be taken to minimize torque and thrust and keep the casing along the design alignment.

6.11 INSPECTION AND MONITORING

The importance of monitoring the work during construction cannot be overstated. Monitoring the progress of the operation enables the construction manager and the inspection team to detect any risks and take mitigating measures in good time. Prior to installation, it is important to inspect the alignment of the track according to the proposed bore path. Before the casing enters the embankment, a thorough inspection should be performed for line and grade. Drastic steering changes cannot be made once the casing is pushed into the ground, but subtle adjustments can be made to ensure that the casing remains on target. This constraint makes it extremely important for the inspection team to be attentive in monitoring the process. The inspection team should also be swift in making decisions that will help the project stay on schedule while meeting the design and quality requirements. The soil condition risks can be minimized by monitoring the excavated soil and comparing it to the geotechnical data used in the bore design. On the basis of this comparative analysis, the construction team can make real, time-appropriate adjustments that can minimize or eradicate these potential risks.

If the jacking or the receiving shaft is very deep, slope inclinometers can be used to monitor the lateral movement of the ground. A slope inclinometer consists of a vertical drill hole with a special grooved casing grouted into place inside the drill hole. An instrument is lowered along the grooves to measure lateral deflection. The information indicates the lateral movement of the excavation, which can be used to evaluate the stability of the shaft excavation (URS Greiner Woodward Clyde 2000).

In cases in which surface settlement or heave is a concern, the surface must be carefully monitored. This monitoring is especially important when there are sensitive structures and utilities nearby. A monitoring plan should be prepared that details the surveying method for, and the frequency of, monitoring the settlement points to detect settlement or heave movements early enough to take corrective actions and prevent or reduce damage. Monitoring during construction is also pivotal to quality management. The contractor should have a quality control plan and should monitor the work and verify that it meets the quality criteria specified in the design documents.

The amount of spoils removed should be checked during and after installation to determine whether voids were created in the soil outside of the casing. Voids would need to be grouted with approved grouting materials to avoid the subsidence of the surface. The possibility of void formation in sandy or unstable soil conditions is high, and grouting the outside of the casing is strongly recommended.

6.12 AS-BUILT DRAWINGS AND DOCUMENTATION

As-built drawings are usually required and should be prepared post-installation for any underground utility construction, including auger boring projects. The contractor should mark the new line, manholes, and ancillary structure information on a copy of the plans to reflect the as-built status. On these plans, the contractor should also document any changes to the original layout of the underground utilities and structures that took place during the construction phase. The rerouting of any utility because of shaft excavation, changing depth, alignment to avoid problematic soil conditions, and so on, should be marked on the as-built drawings. These changes shown on the as-built drawings should be verified and used to update the as-built electronic files for the locality. Line and grade information taken for each drive should be submitted to aid in the development/verification of the as-built drawings.

6.13 RISK ASSESSMENT PLAN

Risk factors that may lead to failure or significantly affect the cost and duration of an auger boring operation include

- Damage to existing utilities.
- Mixed soil conditions.

- Presence of excess water that could flood the machine.
- Hard rock formation.
- Large boulders in the bore path.
- Presence of running sand.
- Excessive bore lengths. With increasing bore length, there is corresponding torque buildup, which at a given length could be excessive and may stall the operation. Increased length also increases the friction forces between the casing and the surrounding soil, which at a given length could exceed the pushing capabilities of the HAB machine and cause problems.
- Drift of casing.
- Slope failures that occur when driving forces exceed resisting forces and could pose danger both to human safety and nearby structures.
- Other safety concerns.

It is important to ensure that the area around every shaft is sufficient for the safe operation of the equipment and for material staging. Acceptable arrangements for traffic control, based on local government regulations, must be carefully planned and followed. If contaminated soil is excavated, the contractor should take the necessary measures to handle and dispose of the contaminated soil. This disposal must be done in conformance with existing applicable regulations, like EPA or NPDES requirements.

6.13.1 Matching System Components to Reduce Risk of Failure

One of the most critical decisions before boring is to ensure that the boring system is compatible with the expected soil and has sufficient power to complete the run from the jacking shaft to the receiving shaft. The system must be able to overcome the friction between the soil and the outside surface of the casing with a reasonable margin of safety. It is also crucial that the boring head configuration be suitable for the expected soil conditions. In soft and running soils, the primary concerns are controlling the intake of the boring head and supporting the overburden soil to avoid settlement. The intake of the boring head is controlled by reducing the opening that allows the soil to flow into the boring head. In harder soil, the challenge is excavating soil with enough speed to complete the run safely and economically. The head may be equipped with cutting teeth and/or cutting disks in the case of rock. Achieving the perfect balance of speed, avoiding settlement, and successfully completing the run is the primary responsibility of the supervisor of the boring contractor.

The contractor should adhere to the sizing guidelines stated in the operations manual issued by the boring system manufacturer to match the system with the needs of the job. The boring system manufacturer should be consulted if there is any doubt about the adequacy of the system for that specific run in those particular conditions (e.g., type of soil, depth, or type of pipe).

6.13.2 Contingency Plan

An important submittal is the contractor's contingency plan. Most contractors have contingency plans that include the corrective actions to be taken if certain events take place. Auger boring projects require the addition of a few specific events to this standard contingency plan. Events that are unique to auger boring and need to be addressed include

- Excessive ground movement,
- Excessive deviation from the target path,
- Slow boring progress or a stuck boring head,
- Damage to an existing utility, and
- Dewatering problems in the jacking or receiving shaft.

Auger boring operations can be performed safely and successfully if the project is well planned, the site conditions are known before the boring operation, and appropriate measures are taken to ensure that the operation is successful.

6.14 DSCs/OBSTRUCTIONS

Regardless of how thorough the geotechnical investigation is, it cannot cover every inch of soil along the auger path. As stated earlier in the design section, the contract documents should allow for adjustments in the project duration and/or cost if unforeseen conditions are encountered. These adjustments should allow for the fair and expedited settlement of change orders. The contractor needs to document the magnitude of the impact in terms of time and cost, along with the cause and effect relationship. The issuance of notices to inform the contracting officer as soon as the problems are encountered is crucial for a successful change order. Most standard contracts recommend that the contractor notify the contracting officer immediately and that the contractor not proceed with the work involved in the differing condition until directed by the contracting officer. Emergencies with safety concerns for the workers or the public may be exempted from this immediate notification on the basis that addressing these emergencies is more critical than the notification. However, the contractor should notify the contracting officer shortly after addressing the emergency.

Obstructions are defined as any buried objects that lie completely or partially within the cross-sectional areas of the planned pipeline excavation that can impede continued forward progress of the auger bore along the design path and within allowable tolerances. Occasionally, a stuck auger bore can be freed by judicious manipulation of the auger, the counter-rotation of the head in quick succession, or the adjustment of jacking pressure. However, when these initial attempts fail to dislodge the obstruction, access to the excavation face is needed to remove or dislodge the obstruction. A 24-in. casing may allow personnel entry to investigate the obstruction; however, the small space limits the ability to solve the problem. Larger casings allow more options to remove the obstruction. Although a 36-in. casing allows much more room than a 24-in. casing, generally a 48-in. casing would be required to enable bore completion by hand mining.

In situations in which numerous cobbles and/or boulders are expected, the contractor may increase the auger bore diameter to improve the ability to deal with cobbles and/or boulders. A larger casing allows better handling of cobbles and boulders through the available higher horsepower and thrust, and access to the face is possible. Most auger boring systems can handle boulders/cobbles that are less than 33% of the casing diameter if the boulders are in a position to enter the casing without stopping the forward progress of the casing and/or affecting the alignment of the casing. Even though it is possible for the HAB system to handle a boulder one-third the diameter of the casing, these boulders can be very problematic to the boring operation. Cobbles (smaller than one-third of the casing diameter or nested cobbles) may stall or damage the auger, but they can pass through the auger shaft. Cobbles larger than 33% may stall and stop the boring process completely. Ideally, manned intervention with a jackhammer or other hand tools can be used to remove these obstructions. Typically, manned interventions would require a casing with a diameter of at least 48 in. (1,200 mm).

6.15 SUGGESTIONS FOR DEALING WITH VARIOUS GROUND CONDITIONS

HAB is a very versatile method that can be used in a wide variety of ground formations. However, the actual soil encountered may differ from the expected soil, and the selected and most fitting system for the expected soil conditions may slow significantly and/or stall. Below are some methods for overcoming these difficulties. The ways of dealing with various subsurface conditions are wide ranging and situation based. Some of the potential solutions described here may assist in dealing with these different ground conditions, as may the methods provided in Table 2-4.

Wet and Running Sand. HAB is generally not advisable in extremely wet running sand, with which there is great potential for settlement. However, if these conditions are encountered, running the cutterhead inside the casing with the wing cutters tucked in may reduce settlement. An overcut will be achieved by the reinforcement band on the leading casing. The rate of soil penetration should be fast while the auger is turned slowly. The pressure should be monitored to avoid excessive jacking force and moderate the speed of the cutterhead. Long bores may require the use of lubricant and continuous installation. The use of a sand auger (an auger section specially built for sandy situations that has a smaller pitch than a regular auger section) is recommended. A high degree of operator skill is required when boring in wet running sand.

Wet and Stable Sand. The cutterhead should be run flush with the end of the casing. An overcut band should be used. The rate of penetration should be fast while the auger is turned at medium speed. Long bores may require the use of lubricant and continuous installation. A steering head with wing cutters may be used in this type of soil. When a steering head is being used, strong lightweight sheet metal can be used to cover the gap where the steering head hinges. The sheet metal should be welded to the front steering section only. A sand auger should be used for insurance against pockets of runny sand.

Dry Sand. If the sand is stable, recessing the cutterhead inside the casing, turning the auger slowly, overcutting, and using a sand auger band are advised. The rate of penetration should be fast while the auger is turned slowly. Long bores may require continuous installation and the use of lubricant. However, in most situations, the main problem with dry sand is the running action of the sand and the subsequent void condition. To prevent this problem, remove the cutterhead and run the lead auger inside the casing about 12 in. (300 mm) or one times the casing diameter, whichever is greater. The use of a steering head is not recommended in this soil condition. A high degree of operator skill is required in this soil condition.

Dry Clay. The cutterhead can be flush or just ahead of the end of the casing. An overcut band is advised. The rate of penetration should be fast while the auger is also turned fast. Long bores may require lubrication. Wing cutters should be used on the cutterhead to provide an overcut. The problem with clay is that it tends to ball up and stick to the turning auger. Water should be added to the inside of the casing to lubricate the clays and get the cuttings out. If the clay is expansive, a hydration inhibitor mixed with the water is recommended to prevent the clay sticking to the auger.

Wet Clay. The cutterhead should be run flush with the end of the casing. An overcut band is recommended. The rate of penetration should be fast while the auger is turned at medium speed. The use of wing cutters is optional, depending on the stiffness of the soil. Long bores may require lubrication. A waterline should be used to add water to the inside of the casing to lubricate the clays and get the cuttings out. If the clay is expansive, a hydration inhibitor mixed with the water is recommended to prevent the clay sticking to the auger. A steering head may be used in wet clays.

Small Gravel. The cutterhead should be just ahead of the end of the casing. An overcut band should be used. The rate of penetration should be medium while the auger is turned at medium speed. Long bores may require lubrication. Wing cutters should be used on the cutterhead to provide an overcut. A steering head may be used in small gravel.

Hard Pan. The cutterhead should be just ahead of the end of the casing. An overcut band should be used. The rate of penetration should be medium while the auger is turned slowly. Long bores may require lubrication. Wing cutters should be used on the cutterhead to provide an overcut. A waterline is used to add water to the inside of the casing to get the cuttings out. A steering head can be used in hard pan. If there is a potential for encountering rocks and boulders in the hard pan, consideration should be given to the casing size. A 30-in.-diameter (750 mm) or larger casing should be used to allow workers to remove boulders that have entered the casing.

Soft/Solid Rock. The cutterhead should be ahead of the end of the casing. A rock head equipped with wing cutters should be used. The auger speed should be medium with an extremely slow rate of penetration. In harder soils, an SBU is recommended. An overcut band and a lubrication line should be used. In the middle of the bore, it may be necessary to replace the teeth on the cutterhead by pulling out all the augers. A high degree of operator skill is required in this soil condition.

Large Gravels/Small Boulders. Traditional HAB is not recommended for this type of installation. Another method, such as hand mining, should be considered.

Hard Rock. For drilling through hard rock, the following methods can be used:

- SBU with disk cutters designed for that rock.
- Drill stem method, which uses roller cone drilling bits to install a pilot hole with drill steel that is then enlarged to the final size with a reaming head. There is no steering control with this method.

6.16 POTENTIAL PROBLEMS AND POSSIBLE SOLUTIONS

The best way of dealing with auger boring problems is to avoid them or reduce their probability of occurrence by properly following the design and construction guidelines in this MOP. However, for cases in which the necessary precautions are not followed or unforeseen conditions occur, this section discusses actions that could help avoid problems, correct them, and/or mitigate their impact. Some of the potential problems associated with auger boring include hitting unforeseen utilities, cutterhead selection, unforeseen obstacles, and site restrictions.

- If excessive ground movement close to an existing structure is anticipated, a ground movement-monitoring plan should be developed. If dangerous movements are observed, stopping the boring is mandated until the causes are analyzed and corrective options are studied (including the option of abandoning the boring method). If a utility is very close to the boring head and at risk of damage, exposing the line may reduce this risk significantly. The excavation to expose this utility should use means that do not damage the line, such as vacuum or manual excavation.
- If cobbles and/or boulders are anticipated along the boring path, increasing the bore diameter will increase the ability of the machine to break and swallow the cobbles and boulders. Cobbles and boulders less than one-third of the casing diameter can sometimes be successfully ejected without the need for worker intervention. In addition, with larger-diameter bores, the augers can be withdrawn to enable safe personnel entry for the removal of the obstructive cobbles and boulders at the face of the excavation.
- The auger boring crew may face a condition in which the cutterhead is not able to cut the soil with sufficient speed because it is boring through harder soil, boulders, or cobbles. If this problem takes place shortly after the start or in the middle of the bore, the solution is to replace the boring head with a more suitable head for the expected soil conditions. The crew can withdraw the augers and the cutterhead, replace the cutterhead with a more suitable head, and then proceed with the bore.
- In some cases, a slow boring rate may be the result of certain components or accessories of the system being undersized or unmatched. Upsizing these components (within the allowable range of that system) is the recommended solution. Matching the system and its components and accessories with the needs of the boring job may solve the problem.
- When boring through soil that is causing excessive friction, the application of a suitable lubricant to the outside surface of the pipe is an effective way to reduce the required jacking force. To apply this solution, a lubrication manifold and lubrication line should be set before the start of boring. This step will enable the lubricant to be pumped during boring.
- It is critical that the contractor ensure that the casing and carrier pipes meet the specifications before, during, and after boring. Adhering to the quality control and quality assurance plans during the manufacturing and shipping of the pipe to the site, along with the

proper unloading of the pipe, reduces risk of pipe failure. It is recommended that the welding of the casing be performed by well-trained welders under appropriate supervision to reduce the risk of failure later, when repair is difficult and costly.

6.17 SURFACE MOVEMENT-MONITORING PROGRAM

The safety of nearby buildings and structures is as paramount in auger boring installations as it is in deep open-cut installations. The safety of nearby structures can be compromised if the structures are subjected to ground movements or vibration that they cannot withstand. When significant ground movement is possible near a critical utility or building, preconstruction surveying and monitoring of the ground movement are advisable. A preconstruction survey of all nearby buildings and structures that documents all existing cracks, cosmetic problems, and structural deficiencies is recommended prior to any work on a site. Preconstruction surveys and elevation monitoring can significantly reduce the risk of damage and lawsuits to the contractor and the owner. The elevations of carefully planned settlement points (on nearby buildings and on the ground surface) around the jacking and receiving shafts should be surveyed prior to boring, during boring, and after boring.

6.18 TYPES OF PAYMENT

The most common method of payment is by unit price per linear foot of casing actually installed. Details can be worked out between the general contractor and the boring subcontractor regarding the scope of work performed. Excavation for the pits is usually included in the cost of the bore. The unit price includes labor, materials, equipment, subcontractors, overhead, and profit.

Another method of payment is by lump sum. Again, the scope of work needs to be clearly determined. Under certain conditions, such as fast-track projects, the cost-plus method (time and materials) can be used.

6.19 COSTS FOR HAB

The cost estimation of HAB projects for bidding purposes needs to be detailed, methodical, and systematic, as is the case for open-cut installations. The project cost consists of direct cost and indirect cost (general expenses and add-ons). The contractor has to estimate the direct cost for labor, material, and equipment for the needed activities for each run. These activities include

- Mobilization;
- Traffic control;
- Clearing and grubbing;
- Pit preparation, including the excavation and shoring of pits, pit bottom stabilization (concrete or gravel), and the installation of the thrust block/plate;
- Potential dewatering;
- Preparation and installation of steel casing;
- Installation of carrier pipe and grouting of the annular space;
- Site restoration, including backfilling and compacting, seeding and mulching, and cleanup; and
- Demobilization.

Each of these activities (cost items) may be broken down into subitems or subactivities, for which the labor, materials, and equipment need to be estimated. Other general expenses (any expenses that can be estimated independently of direct cost expenses) and add-ons (usually added as a percentage of other expenses, such as labor overhead) need to be estimated and added to the direct cost. The final bidding price is estimated on the basis of the bidding strategy of the contractor.

6.20 SAFETY ISSUES

All the standard safety procedures followed in typical open-cut construction are applicable to auger boring projects. Additionally, the workers should understand the components and operation of the HAB machine, particularly the moving parts in the system. The HAB machine should thrust against a thrust block and/or the anchoring pins that (along with the soil behind them) should withstand the jacking forces. The stability of the soil behind the thrust block and/or the anchoring pins should not be compromised. Prior to boring, the contractor has to ensure that there are no unforeseen gas, water, wastewater, power, or fiber optic lines close to the boring head.

Safety on the jobsite is the responsibility of the contractor. For safe operation, the contractor must follow all the applicable OSHA standards and manufacturers' safety/operations instructions. Inexperienced and untrained personnel should not be allowed to operate the equipment or work near the machine without careful supervision by a qualified person. It is critical to ensure that all personnel are well trained to perform the boring operations safely. Personal protective equipment, including hard hats, steel toe boots, and eye protection, should be worn on site at all times. The appropriate eye protection should be used for welding operations and all other operations that require protection.

For welding operations, the ANSI Z49.1:2005, *Safety in Welding, Cutting, and Allied Processes* (AWS 2005), provides that

- All cutting equipment shall be inspected as required to ensure that it is in safe operating condition. When found to be incapable of reliable safe operation, the equipment shall be repaired by qualified personnel prior to its next use or withdrawn from service.
- All equipment shall be operated in accordance with manufacturers' recommendations, provided they are consistent with the standards.
- Managers shall ensure that welders and supervisors are well trained in the safe operation of the equipment, the safe use of the process, and emergency procedures.
- Workers or other persons adjacent to the welding areas shall be protected from the radiant energy and splatter of welding and cutting by noncombustible or flame-resistant screens or shields, or shall be required to wear eye and face protection and protective clothing (AWS 2005).

6.21 CONTROL OF CONSTRUCTION NUISANCES

Noise. The best way to minimize this nuisance along with other construction nuisances is through source controls. The control of noise may entail using newer and quieter equipment that is able to work at lower power levels and that has advanced mufflers and sound insulators (Peurifoy et al. 2011).

Dust. The contractor should prepare a dust control plan, especially when working in an urban setting where the dust might generate complaints. Dust control could entail prewetting, watering, and the use of coverings and windscreens.

REFERENCES

Atalah, A., Iseley, T., and Bennett, D. (1994). "Estimating the required jacking force." *Proc., No-Dig 94 Conf.*, NASTT, Dallas, TX.

AWS (American Welding Society). (2005). "Safety in welding, cutting, and allied processes." *ANSI Z49.1*, Miami, FL, 3–5.

Iowa DOT. (2009). "Casing pipe." <http://www.iowadot.gov/design/dmanual/11D-01.pdf> (Jun. 28, 2016).

Midwest Mole. (2014). "Auger boring." <http://www.midwestmole.com/images/auger-boring5.jpg>.

Najafi, M. (2013). *Trenchless technology: Planning, equipment, and methods*, McGraw-Hill, New York.

Permalok. (2002). "Permalok. The interlocking pipe joining system." <http://nwpipelok13.alpineclients.com/t7_test1.html> (Feb. 15, 2014).

Peurifoy, R. L., Schexnayder, C. J., Shapira, A., and Schmitt, R. L. (2011). *Construction planning, equipment, and methods*, McGraw Hill, New York.

URS Greiner Woodward Clyde. (2000). "Microtunneling design guidelines." Offshore and Onshore Design Applications Supervisory Committee of Pipeline Research Council International, Inc., Houston.

CHAPTER 7
CONCLUSIONS

The HAB method is generally used to install steel casing pipes under railways and roadways and above the water table in relatively soft, stable ground conditions, such as clay, silt, or sand with clay and/or silt in the soil matrix. The method can accommodate cobbles and boulders up to about 33% of the casing diameter. Recent innovations have extended the capabilities of the HAB machine to cut through medium to hard rock conditions and mixed ground conditions with relatively high precision. The SBU is a small-diameter cutterhead and thrust-bearing assembly that extends the capabilities of the HAB machine to cut through medium to hard rock conditions and mixed ground conditions. The SBU has been used to cut rock exceeding 25,000 psi (170 MPa) and can be used for installations ranging from 24 to 78 in. (600 to 2,000 mm) in diameter. The PTM, which was first introduced in the United States in 1995, makes it possible to achieve a greater level of accuracy [±0.25 in. for 300 ft (±6 mm for 90 m)]. The PTM is used to install concrete, clay, polycrete, and steel pipe ranging from 4 to 48 in. (100 to 1,200 mm) in diameter with a drive length of up to about 300 ft (90 m).

The more recent innovations that have both extended the capabilities of HAB and made the system safer include the CBS, the steel pipe interlocking joining system, the laser-guided tunnel attachment, the on-target steering system, the BMTA, the use of the GBM to install pilot tubes that the HAB method can follow, and several safety enhancements.

The planning phase of an auger boring project is critical because it outlines the basic elements required for successful completion. Planning would also involve examining and determining the suitability of HAB for the project. This phase provides sufficient information to allow the design engineer to develop detailed design drawings and contract documents that will be used by the contractors to estimate the cost of the project competitively with appropriate risk allocation. A well thought out and properly executed geotechnical investigation program provides data of great value to project success. Geotechnical considerations alone can affect the feasibility of almost any type of project, and HAB projects are no exception. The geotechnical investigation and analysis guide all parties to identify the inherent risks of underground construction. The magnitude and depth of a geotechnical investigation depend on the risks associated with the project.

The owner's design engineer is responsible for the long-term design and the specification of performance criteria for the contractor's means and methods. Accurate and detailed design documents prepared by the owner's engineer will permit a qualified contractor to complete the construction successfully. The contractor, however, is in charge of the construction design, which includes the support of excavation for the jacking and receiving pit, the dewatering system, the jacking equipment, the thrust blocks or other reaction system necessary for the jacking equipment, the casing to resist jacking forces and potential misalignments, the spoils management and disposal conditions, and the annular fill design.

When good design and construction guidelines, including those presented in this MOP, are followed, the probability of occurrence of the various risks and their impact on project cost, quality, and schedule are significantly reduced.

GLOSSARY

Advance—The motion of the boring machine toward the face wall of the entrance pit.

Annular space—The space between the borehole and the installed casing pipe or the space between the casing and the carrier pipe.

Auger—A helically wound drive shaft, with hex couplings at each end, that transmits torque to the cutterhead and transfers spoils to the drive pit.

Auger drive or drive chuck—The female hex connector located within the casing pusher.

Auger string—A set of augers connected end to end to extend from the cutterhead to the HAB machine.

Backstop—Reinforced area of the boring pit wall directly behind the track.

Band—A ring of steel welded at or near the front end of the lead section of casing to cut relief and strengthen the casing.

Bedrock—A general term for the rock, usually solid, that underlies soil or other unconsolidated bed material.

Bentonite—A colloidal clay, sold under various trade names, that forms a slick slurry or gel when water is added.

Bore—A generally horizontal hole produced underground, primarily to install services.

Boring—The dislodging or displacement of spoils by a rotating auger or drill string to produce a hole called a bore.

Boring head—See "cutterhead."

Boring machine—A machine to bore earth.

Boring pit or entrance pit—An opening in the earth of specified length and width in which the machine is placed on line and grade.

Carriage—The mechanical part of a non-split boring machine that includes the engine or drive motor, the drive train, the thrust block, and the hydraulic cylinders.

Carrier pipe—The pipe that carries the product being transported (water, gas, sewage, or other products and services). The carrier pipe may be installed within a casing when crossing a highway or railroad. The pipe may be made of steel, concrete, clay, glass fiber-reinforced polyester, plastic, ductile iron, or other materials. Some carrier pipes can be installed directly using other installation methods.

Cased bore—A bore in which a pipe, usually a steel sleeve, is inserted simultaneously with the boring operation. This term is mainly associated with auger boring.

Casing adapter—A circular mechanism to provide axial and lateral support to a casing whose diameter is smaller than that of the casing pusher.

Casing attachment—See "Casing adapter."

Casing pipe—A direct-jacked pipe in a two-pass system that supports a bore. The casing is not a carrier pipe or a product pipe; such pipes are installed within the casing pipe.

Casing pusher—The front section of a boring machine that distributes the thrusting force of the hydraulic cylinders to the casing and forms the outside of the spoils ejector system.

Centerline—The vertical distance between the center of the drive chuck and the ground plane (bottom of the track).

Cleaning—An action of a boring machine to remove spoils that occur when the auger is rotating while axially stationary.

Clutch—A mechanical device that engages or disengages rotary torque from a power source.

Collaring—The initial entry of the casing or a cutterhead into the earth.

Conduit—A broad term that can include pipe, casing, tunnels, ducts, and channels. The term is so broad that it should not be used as a technical term in boring or tunneling.

Control lever—A handle that activates or deactivates a boring machine function.

Cross members—The lateral supports under the track of a boring machine.

Cutterhead—Any rotating tool or system of tools on a common support that excavates at the face of the bore; usually applies to the mechanical methods of excavation (Najafi 2013).

Deadman—A fixed anchor point used in advancing a saddle- or cradle-type boring machine.

Deck assembly—Drive train assembly for a split design boring machine.

Dewatering—The use of a system of wells and pumps to remove groundwater or lower the water table.

Dogs—Movable protrusions in the thrust block that engage holes or blocks in the track.

Emergency controls—Those controls that stop power to the machine or its components.

Emergency stop—A red manually operated push button that, when activated, stops all functions of the machine.

Exit pit—An excavation located at the exit of the cutterhead or casing.

Extension track—An additional section of track used in front of the master track.

Face—The location where the excavation is taking place.

Flight—The spiral plates surrounding the tube of an auger.

Forward rotation—The clockwise rotation of the auger as viewed from the machine end.

Grade—The specified rise or fall of the proposed bore from a horizontal plane. Calculated as rise/run × 100%.

Horizontal auger boring (HAB)—A technique for forming a bore from a drive pit to a reception pit by means of a rotating cutterhead. The casings are jacked forward sequentially in a cyclic process while the auger is turned. The spoils are moved back to the drive shaft by the rotation of the helically wound auger flights in the steel casing. The equipment may have limited guidance and steering capability.

HAB machine—A machine used to excavate the ground horizontally by means of a cutterhead and auger or other functionally similar device. The machine may be either cradle or track type.

Jacking lug—See "Deadman."

Product pipe—See "Carrier pipe."

Restoration—The backfilling, compaction, and resurfacing of any excavation to restore the surface and underlying structures.

Retract—The motion of the machine away from the face of the entrance pit.

Reverse—The counterclockwise rotation of the auger as viewed from the machine end.

Saddle—A vertical support mechanism to hold the casing in position while starting (collaring) the bore.

Saddle adapter—An attachment that fits on the saddle to support a smaller casing.

Safety sign—A notice attached to the machine that advises about the nature and severity of a potential hazard that can cause injury or death. The sign can also provide instructions to reduce or eliminate the hazard.

Shank—A hardened male hex bar containing one or more transverse holes to couple and hold in a female hex connector.

Shield—A guard that alone or with other parts of the machine provides hazard protection from the area covered.

Skin friction—Resistance to the thrust caused by earth pressure around the casing.

Small boring unit (SBU)—A small-diameter cutterhead and thrust-bearing assembly that is used in concert with a HAB machine to enable effective boring through medium to hard rock conditions and mixed ground conditions.

Sonde—A transmitter that emits a continuous magnetic signal at a set frequency. The receiver displays the signal sent by the transmitter.

Split design—A boring machine that can be broken down into two or more sections to reduce the lifting weight.

Spoils—Earth, rock, and other materials excavated during the installation process. Also referred to as *cuttings*.

Spoils ejector—A set of paddles rotating in close proximity to the inside of the casing pusher.

Spoils ejector door—A safety guard that partially or completely covers the spoils opening in the casing pusher.

Steering head—A movable lead section of casing that can be adjusted to steer the bore.

Thrust—The force that causes the boring machine to advance.

Thrust bearing—An external bearing used to isolate the final drive from the thrusting force of the machine.

Thrust block—An engineered structure, located between the jacking frame and the shaft wall, which distributes the jacking force developed by the hydraulic jacking frame over a large surface area to the ground behind the back wall of the shaft.

Thrust package—The bottom section of a split boring machine. The thrust package contains the cylinders and the thrust block.

Torque—The measure of the rotary force available at the drive chuck.

Torque limiter—A rotary slip clutch used to protect the final drive.

Torque plate—Steel plates welded to the casing and bolted to the casing adapter to prevent the casing from slipping.

Torque windup—The energy stored when metal parts are held in a twisted condition. Release of this energy can cause the unexpected rotation of the auger.

Track—A set of longitudinal rails mounted on cross members that support and guide a boring machine.

Track brake—A mechanical device to provide a limited resistance to movement between the machine and the track.

Track pins—Steel pins to be driven through holes in the track into the base of the pit to hold the track in place.

Transmission—A gear reduction unit located between the power source and the final drive.

Trench box—A preconstructed set of side plates and adjustable cross members to prevent the walls of the pit from collapsing.

Trenchless technology—A family of construction techniques for installing, repairing, replacing, or rehabilitating underground infrastructure with minimal disruption to traffic, businesses, and residents. The grouping also includes technologies for inspection, leak location, and leak detection with minimal disruption and minimal excavation from the ground surface.

Underground utility—Active or inactive services or utilities already in the ground in the area of the proposed bore.

Upset—The inadvertent action of a boring machine that rotates the machine and track from their normal upright position to another position.

Water level—An instrument that uses a tube filled with water to indicate the elevation of the lead section of casing.

Water table—The elevation of the groundwater.

Wing cutters—Appendages on the cutterhead that open to increase the cutting diameter of the head when turned in a forward direction and close when turned in a reverse direction. They are used to cut clearance for the casing pipe.

REFERENCES

Najafi, M. (2013). *Trenchless technology: Planning, equipment, and methods*, McGraw-Hill, New York.

INDEX

Page numbers followed by *f* and *t* indicate figures and tables, respectively.